Math Integrity Understanding Strong and Weak Force Through the Forces / Fields of Electrostatic, Direct and Axial Nucleostaticmagnetics: 4-Vector in 3D Model Generating Four Quantum Equations (Dirac)

By Arno Vigen

One of the most challenging concepts in physics of the last century is **strong nuclear** force/interaction and **weak nuclear** force/interaction. Each of those forces is used often, but what meaningless names. Strong versus what? Weak versus what? Are we stuck in this abstract, disconnected model? No.

I postulate that those two fundamental forces are really **the static equations and underlying concept of magnetism at the subatomic particle level**. In the creation of electrostatic charge for subatomic particles, that activity also creates a subatomic particle axis (I call "nucleostaticmagnetics" or "N-M") which generates both a static towards-the-axis force (weak), and an outward bulge (strong) repulsive force (electron-nucleon). Both work at 1/distance-cubed ($1/d^3$) magnetic strength (so only 'nuclear').

These two static forces create a constant conflict by those two very different orientations. Yet also, this defined set of forces work together, as electro- plus -magnetic, to create all the observed behaviors that are the properties of Elements of the Periodic Table, quantum behaviors, and magnetism.

Now, the current naming, that almost everyone dislikes, needs to get updated. First, these historical names are not useful, and the replacement with 3D engineering concepts of **direct** (in the line between the particles) and **axial** (towards-the-axis) will make these easy to understand. Second, the two **'nuclear'** forces today **are each a combination**, as I shall explain, that confused their application, and made the underlying concept difficult to understand. When I unpack each of these systems into four (4) vectors, with a 5th time-dependent reactive vector, then the system will lead to better 3D engineering deterministic and continuous from the subatomic level to the macro-world level.

Of course, it starts with isotropic **electrostatic (Coulomb)**, which remains steady in all applications here. After that, to engineer the postulated system, the two 'nuclear' forces calculate by three more vectors:

- A ('strong') vector for electron-nucleus provides a physics-positive (repulsive) force directly towards the other electron-nucleus particle-set (yet, it is physics-negative attractive between nucleons within the nucleus given the herein proton-neutron-proton 3D nucleus structure requirement[i]).

  - A ('weak') vector is:
    - for direction, as the combination of:
      - a vector pointed towards the nucleostaticmagnetics axis
      - the tangent plane relative to the electrostatic same-strength sphere (isotropic).
    - for strength, based upon $\sin(\theta)*\cos(\theta)$[ii] of the 'strong' vector as a force integral over the triangle defined by the nucleus and the 90-degree drop from particle to nucleus axis[iii]

In the drawing below for electron-nucleus interaction, direct electrostatic force (red solid arrow) is attractive, but that is offset by the repulsive direct nucleostaticmagnetics (strong nuclear) force (green solid arrow). Both are along the particle-particle orientation. Those are the solid line vectors below.

Note the engineering is 3D, but I present in as only the paper 2D slice here.[iv] 3D parts will come later.

However, each of those has a field. Yet, those fields act quite differently. The nucleostaticmagnetics (strong nuclear) force generates a vector (green hashed arrow) back towards the closest position of the nucleostaticmagnetics axis (purple double hashed line) of the other particle. Notice that is a direct drop, so 90 degrees to that axis. That is the preferred movement direction if that force vector alone.

However, the field of the axial electrostatic force is isotropic, so the same strength is a huge sphere (red dotted circle), and that has a tangent (red hashed arrow) to that sphere for movement preference.

That is two movement choices; and that means the favored path for movement is the sum of those vectors (geometry averaging) which is a vector at ~θ/2 off the perpendicular to the axis (the brown hashed arrow). In many ways, but not all, as I will explain, that makes certain (but not other!) calc's operate as if a regular toroid (brown dotted circle) fixed by the nucleostaticmagnetics axis (purple hashed line). So, below describes how those vectors (force directions) change relative to each other:

(01)

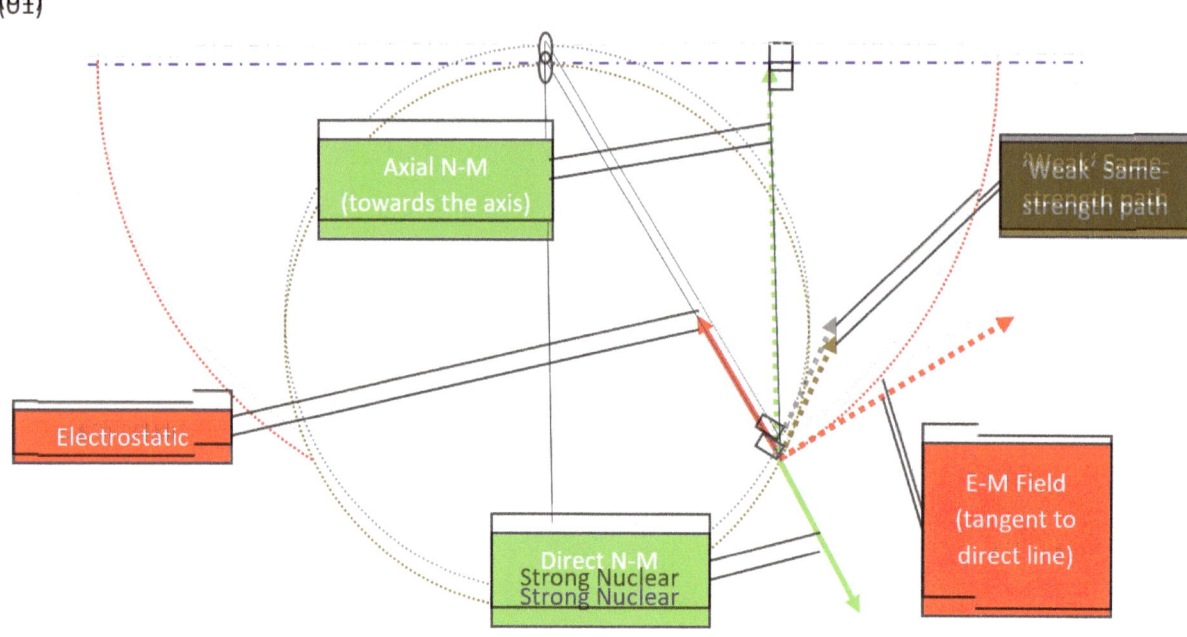

That wording 'strong' and 'weak' needs to get replaced with **direct** (strong nuclear) and **axial** forces, with 'weak' direction in movement that average net-2-field-weak field path, as in combining a) (NM-field-vector) towards-the-axis, with b) the (ES-field-vector) same strength electrostatic ("E-S") field. So, weak is the net of 2 vectors for direction. These forces become vectors that we can 3D engineer.

As one can already see, these vectors describing the forces include: a) forces directly along the object/particle line, as well a b) the vector of ES-field as tangent (so orthogonal) to that a) line, but also, c) the towards-the-pole force that is neither a) nor b). Add, the right-hand, and that makes the 3D engineering very complex. So, this process requires reviewing both particles, forces, and fields with that additional layer of preferences to the same-field strength that is critical to understanding magnets.

Now, the **traditional magnetic force** is not quite the above; a traditional magnet uses **a time-dependent added force** about the change in these fields. That has the extra complexity of change in time, which provides a) in the common magnetic field generation occurs as the change in the above fields. Importantly, a magnet is a huge change from both-repulsive poles in the static view to north-south with one just emptied and thereby physics-negative attractive, and the other just filled, so physics-positive repulsive; b) that right-hand rule; and c) aggregation from multiple molecules then generate aggregating or interfering patterns (which themselves interact). That means learning an additional time-dependent force, not just the **static fundamental forces**.

As part of that reactive force, the **Lorentz effect** time-dependent engineering requires that the field propagation is at the relativity speed of light for the particles and fields involved, and that generates a further effect as a counterforce for field strength change, much like the above traditional magnetism counter force. That also is beyond the static model. Those concepts come much later.

Further, there is force directions, but this vector addition/averaging for particle movement direction. And that movement, generates another responsive counter force (Lorentz). That is lot.

There are many steps, so we will start with only the static model, and build from there.

First, the static model creates the basics of 'preference for settling at the poles first' with the first two electrons (subshell-s) in each shell as closer, and all Elements and properties of the Periodic Table by my Scrunched Cube Atomic Model for other layers. It gives the physical properties for all Elements.

Second, the full postulate restores **math integrity**. Direct nucleostaticmagnetics (strong nuclear) force and axial nucleostaticmagnetics (as or leading to weak nuclear) are **not start-and-stop forces**. The forces apply at every distance with the same force equations. This replaces the prior teaching that they (strong and weak force/interactions) are only in certain ranges. Strong force is not just in the nucleus, but also with electrons creating the shells/fields/layers. Weak force is not just in the nucleus, but also with electrons creating the 3D engineering for the other subshells.

So, third, these postulates make both the direct nucleostaticmagnetics (strong nuclear) force and axial nucleostaticmagnetics (weak nuclear) force operates not only inside the nucleus to create nucleus binding, **but also** with the electrons to generate electron shell and subshell structure in a continuous function. The **same equations/forces work at all distances**; it only depends on the particles and distances involved.

Finally, the postulated combination of three static fundamental forces [electrostatic; direct nucleostaticmagnetics (strong nuclear), and axial nucleostaticmagnetics (weak nuclear)] a) are fully interlinked; and b) operate as a set. For example:

- Electrostatic force has an equilibrium with direct nucleostatic to create a) nucleus binding for proton/neutron 3D structures and b) electrons shells/fields for electron/nucleon interactions

- Direct electrostatic (strong nuclear) force works with axial electrostatic (weak nuclear) force to create all the attributes of magnetics (when connected to the time-dependent model/Lorentz)

The three fundamental forces (or four when adding the directional vector of the E-S field/tangent or five with the time-dependent forces) operates as an interlinked set, especially within the atomic range. That limited range, called 'nuclear' for two of them, is because two of the forces decline exponentially more rapidly. By the basic math of 1/distance-cubed ($1/d^3$), these forces are immaterial beyond the distance range of one atom. That means the nuclear force exist at all distances, but get ignored properly beyond the subatomic distance. Math integrity.

So, the definitions of the forces are:

- Electrostatic force is isotropic at 1/distance-squared ($1/d^2$) in the direction between the two particles (Coulomb)

- Direct nucleostaticmagnetics ('strong nuclear') force strength decrease at 1/distance-cubed ($1/d^3$) in the direction between the two particles, so also isotropic.

- Axial nucleostaticmagnetics (weak nuclear) force has strength decrease at 1/distance-cubed ($1/d^3$) **in the direction of the axis** of the other particle with its related field.

- However, that N-M field also must integrate with the **electrostatic spherical field**, so the tangent to that remains the same-strength preference.

- That makes the net-weak force direction, the balancing of those directions, so θ/2, halfway between those two (the axis versus the E-S spherical tangent). It is not just anisotropic, but towards-the-axis vector direction different than either of the two causes. This θ/2 is not direct line between the particles, and not perpendicular to that line. It changes with θ. The direction of that axial nucleostaticmagnetics (weak nuclear) force is moved θ/2 to balance with the easiest same-strength field of basic electrostatic. That is the lines of same strength, perpendicular to the particle-particle direction, move the direction of axial nucleostaticmagnetics calculated for strength above.

- The axial forces also create a set of rotational forces (wave functions) on the particle to rotate its axis to align with the other particle's (the strongest) position which become the 3D engineering of the wave function in quantum mechanics.

- Further, the inclination angle generates a reduction in strength of direct to axis N-M based upon sine times cosine ($\sin(\theta)*\cos(\theta)$). This follows from the ratio of the axis in the same hemisphere providing force, so limits to 0.5000. The axial strength is weaker than direct force by that factor.

- All of the above becomes a set of physics battles: a) between direct isotropic forces and towards-the-poles ones; b) between 1/distance-squared versus 1/distance-cubed; c) between linear forces and rotational forces (to discuss much later), and d) linear forces versus rotational forces, and e) finally between sets of particles as chaos or harmonics.

Forces Magnitude – Physics-Positive (Repulsive) or Physics-Negative (Attractive)

Electrostatic forces are based upon Coulomb's Law which is the multiplication of the charges of the two interacting particles. Opposites attract. Like-kind repel. The rules for interactions between particles for electrostatic force is based upon multiplication, and the sign of the charge each particle multiplied. Opposites multiple to be physics-negative attractive. Like multiple to always be physics-positive repulsive:

(02)

	Proton (+)	Neutron (0)	Electron (-)
Proton (+)	+ * + = + Physics-repulsive	+ * 0 = 0 No interaction	+ * - = - Physics-attractive
Neutron (0)	0 * + = 0 No interaction	0 * 0 = 0 No interaction	- * 0 = 0 No interaction
Electron (-)	- * + = - Physics-attractive	- * 0 = 0 No interaction	- * - = + Physics-repulsive

The nucleostaticmagnetics (strong/weak) rules for interactions are quite different. The rules are somewhat reverse of electrostatic; well, it is a Newtonian counter-force in my mind. Further, both in nucleostaticmagnetics, both protons and neutrons act the same. In the end, it really is two classes:

- Nucleons (protons or neutrons) interact with each other as attractive (physics-negative)
- Electrons interact with nucleons as repulsive (physics-positive)
- Electrons do not interact for nucleostaticmagnetics with each other.

The rules of nucleostaticmagnetics is really about the nucleons (proton/neutrons). It is attractive to other nucleons, and repulsive to electrons, with no interaction between electrons. Now, since the electrostatic presented at a 3 x 3 table, I present the rules for nucleostaticmagnetics with the same matrix. However, the interactions are not from the particle charge (+) to (-), but based upon the below:

(03)

	Proton	Neutron	Electron
Proton	(-) Attractive	(-) Attractive	(+) Repulsive
Neutron	(-) Attractive	(-) Attractive	(+) Repulsive
Electron	(+) Repulsive	(+) Repulsive	No interaction

The electrons not interating is not a complete answer. Since they interact with nucleons, electrons do have a anisotropic (varying) field, and thereby do by rotation generate wave functions, but that is only towards the nucleus. The electron-electron is the one complete interactions as electrostatic-only.

The main factors to understand in physics in particle-level 3-force interactions applicable here are:

- Electrostatic force is 1/distance-squared ($1/d^2$), and a magnet force is 1/distance-cubed ($1/d^3$)
- These two magnet-like forces get calculated based upon nucleons (protons + neutrons)*= mass as used for macro-magnets.   * *not just the protons of electrostatic charge force*
- Magnet-like force has that strange shape of tight at the two poles with bulging at the equator

Next, I find it easier to learn to calculate the absolute-value **magnitude**. Then we will apply the plus or minus as in the section above. And for that we have two basic formulae:

- Electrostatic (Coulomb) magnitude at 1/distance-squared ($1/d^2$) based upon the charged particles (protons (+) and electrons (-))

(04)

$$F = \frac{kQ_1Q_2}{[(x_1-x_2)^2 + (y_1-y_2)^2 + (z_1-z_2)^2]}$$

The product Q_1Q_2 generates the Table (2) logic as a multiplication, so the physics-positive or physics-negative work explicitly from the Coulomb formula. That is why electrostatic (Coulomb) is the easiest and most understood. Note that the Table (02) positive/negative is based into the Q_1 times Q_2:

The same is not true for the postulated Nucleostaticmagnetics force equation. It needs the more complex Table (03) work to get the physics-positive (repulsive) or physics-negative (attractive) with the equation that the total nucleostaticmagnetics force (M_1 or M_2) of each particle is always positive:

- Direct Nucleostaticmagnetics (this postulate) magnitude at 1/distance-squared ($1/d^3$) based upon magnetic particles (protons, neutrons, and electrons)

(05)

$$\frac{M_1 M_2 * \pm Table(03)}{[(x_1-x_2)^2 + (y_1-y_2)^2 + (z_1-z_2)^2]^{\frac{3}{2}}}$$

The 3/2 factor is the ½ for the square root to get distance times the magnetic reduction factor of 1/distance-cubed ($1/d^3$):

Math Logic for Magnetic Force Strength at 1/Distance-Cubed (1/d³)

In fact, I think understanding would improve to spend a moment introducing by 1/distance-squared and 1/distance-cubed apply to force. It is a path to the physical world. It is based upon three dimensions in space and the extra dimension for reduction along the magnetic axis.

An isotropic force, like electrostatic, decreases the same in every direction. That means that the same force at speed of light time later, not covered a larger space. Yet, by conversation, the total force must be the same. As such, the field strength reduces by distance-cubed. (Wait, you thought electrostatic force decreased 1/distance-squared?) However, the observer gets all the force from one dimension. So, in that sense, the observer also has all the strengths available in that dimension. They aggregate (math integral). That means the reduction is only the two dimensions that do not align with the observer.

In the simplest sense, when off the axis, there is a block of space that is defined by two dimensions. One dimension along the axis (in my work z-), and the other dimension off the axis (in this work x,y-). If adding (math integral) from any place, there is a force to every point along that axis as 1/distance-squared, but with that extra direction, the axis, as a measure of reduction. So, the math integral is:

(06)

$$S \equiv \frac{-\frac{1}{2}kQ_1Q_2}{d^3}$$

$$F \equiv \int_{observer\ direction=-infinity}^{observer\ direction=+infinity} \frac{\frac{1}{2}kQ_1Q_2}{d^3} \equiv -2\left(\frac{-\frac{1}{2}kQ_1Q_2}{d^2}\right) = \frac{kQ_1Q_2}{d^2}$$

Now, most people do not teach electrostatic force as a) the integral of field strengths for an observer in one dimension or b) as fields with that extra -1/2 factor. It is simpler to jump to the force, and avoid taking the time to work the integral, especially with the easy spherical force of electrostatic.

However, that is important to the understanding of magnetics. As much as the above is that easy spherical shape. Further, that makes the physical three dimensions have a direct link, by one (1) dimensions math integral, to get the electrostatic 1/distance-squared (1/d²).

So, the electrostatic force in expanded physics form (x,y,z) is a matrix. However, the addition is simple for isotropic force. The magnitude (kQQ) remain unchanged for any interactions. However, the distance does change. So, in that way, we look at the equations

Of course, to get the towards-the-axis, I will need to work from electrostatic as a basic 1/distance-squared vector.

$$F = \begin{matrix} \dfrac{x}{d} * \dfrac{kQ_1Q_2}{d^2} & \dfrac{xkQ_1Q_2}{d^3} \\ \dfrac{x}{d} * \dfrac{kQ_1Q_2}{d^2} = \dfrac{ykQ_1Q_2}{d^3} \\ \dfrac{z}{d} * \dfrac{kQ_1Q_2}{d^2} & \dfrac{zkQ_1Q_2}{d^3} \end{matrix}$$

(07)

Again, one can see how that gets back to the 1/distance-cubed of 3D space. Further, one can see that the total magnitude force gets back to the same think because of Pythagorean theorem.

(08)

$$\sqrt{\dfrac{x^2}{d^2} + \dfrac{y^2}{d^2} + \dfrac{z^2}{d^2}} = \sqrt{\dfrac{x^2 + y^2 + z^2}{d^2}} = \sqrt{\dfrac{d^2}{d^2}} = \sqrt{1} = 1$$

The comparison for a magnetic field is the field strength to all the axis positions (x,y,z so three dimensions), but also each calculation then is relative to the direction of that axis.

Better understood by a graphic. The accumulation of force is field from every point on the magnetic axis. All points exert a field pulling the particle.

(09)

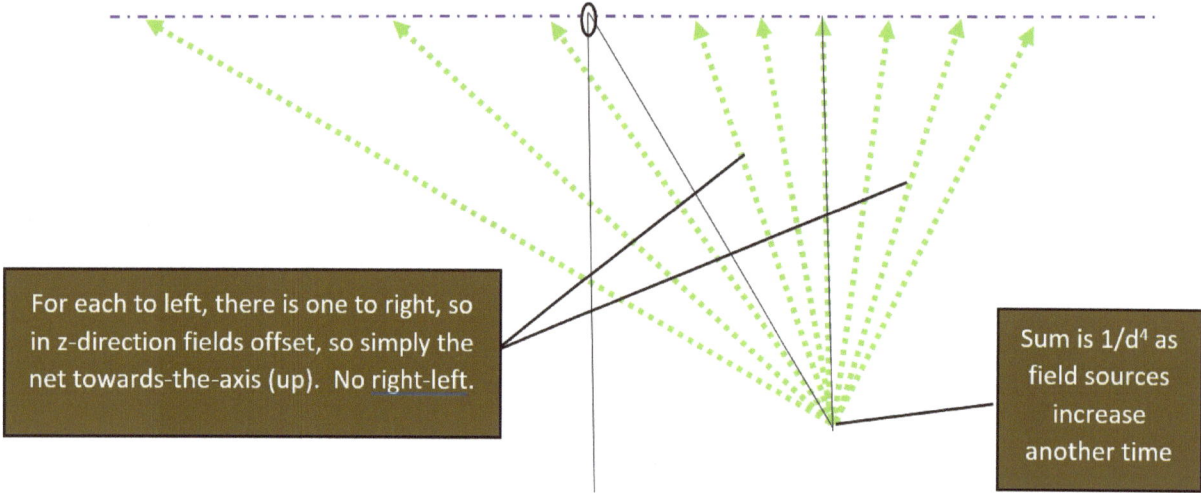

For each to left, there is one to right, so in z-direction fields offset, so simply the net towards-the-axis (up). No right-left.

Sum is $1/d^4$ as field sources increase another time

So, classical equations are just a vector directly towards the axis. That is true because the left portions and the right portions fully offset (with the coming exception that I will explain).

(10)

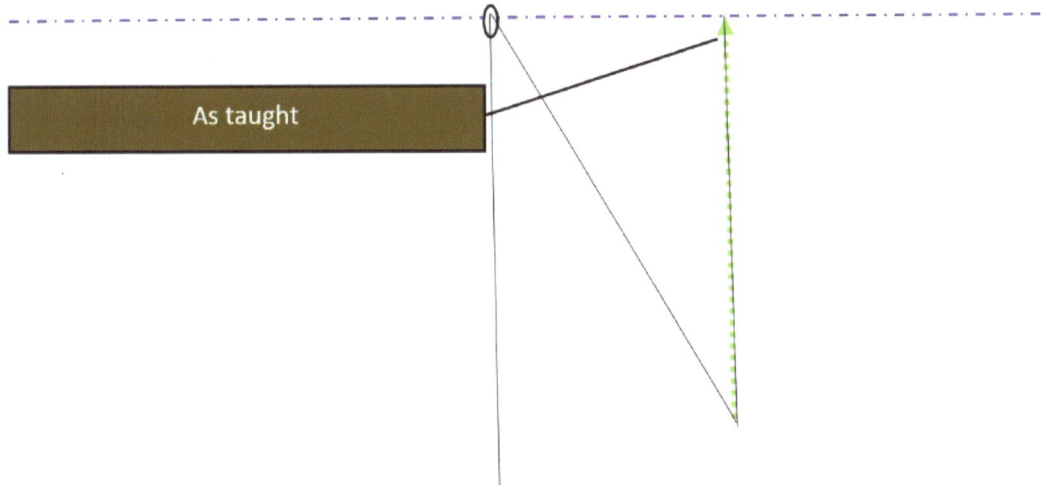

As taught

Once again, must teaching and equations ignore the field, ignore the z-dimension directions that offset. For 99% of challenges, that works great and takes much less time.

In fact, it works great for Maxwell, Biot-Savart for aggregated movement in a loop induced magnetic fields, <u>but only if one is more than say 10x away from the loop itself</u>. Inside that range, the Biot-Savrt does not work. That is because of those darn left and right segments will become important. That is why I am taking the time to introduce them here – when not required.

One has a field which is from all the axis positions, each at 1/distance-cubed. Yet, those field strength are decreasing, by the distance increasing, also, the magnetic fields are every distances, so an added dimension for reduction. In that sense, the total field strength is now reducing by four factors, the spread of each basic field in three dimensions, plus the reduction of all the sources in the fourth factor, their increasing distance relative position along the axis for the many sources.

So, now one gets the integral of field strength $1/d^4$ which becomes total force of 1/distance-cubed ($1/d^3$) for all magnet-like equations. The force gets reduced by the basic 3D plus the 4th axis dimension.

(11)

$$S = -\frac{\frac{1}{3}M_1M_2}{d^4}$$

$$F = \int_{m-axis=-infinity}^{m-axis=+infinity} -\frac{\frac{1}{3}M_1M_2}{d^4} dm = -3\left(-\frac{\frac{1}{3}M_1M_2}{d^3}\right) = \frac{M_1M_2}{d^3}$$

Do not worry about the Q1 versus the M1. Those are the related and are constants in the sense that particle-sets do not change their electrostatic charge (Q by the number of charged particle - protons or electrons) or inherent nucleostaticmagnetics strength (M by the number of nucleons or electrons).

Instead, for this section, the focus is on the field reduction, the denominator, which is the factor which gets integrated from all the field strengths into the overall force on the particle.

Magnetic forces are all calculated based upon 1/distance-cubed based the fields reducing based upon an extra dimension of reduction before the math integral in the object direction to get the total effective force. However, that reduction gets nuanced by whether a) one is far along that axis, or b) within the distance that must calculate net-2-force or net-4-force with both electrostatic and nucleostaticmagnetics.

Of course, that all depends on the relative angle of the observer (electron in this case), relative to the axis with the nucleus as a vertex. That introduces the sine and cosine function. The field strength at the peak is not generally the electrostatic field strength.

That is, the start of the field strength is reduced by that sine function.

The full strength force would be based upon the angle of 90 degrees. The drop is sine(90) which is 1.000. As such, the base strength to start all magnetic fields with the sine function.

The strength at the equator (90-degrees) grows to about 1x.

(13)

$\cos(\theta)$

$\theta > 90$

$\sin(\theta)$

At other inclination angles, the strength starts at strength based upon that angle. If one know, then the calculation is easy using the full strength.

(13)

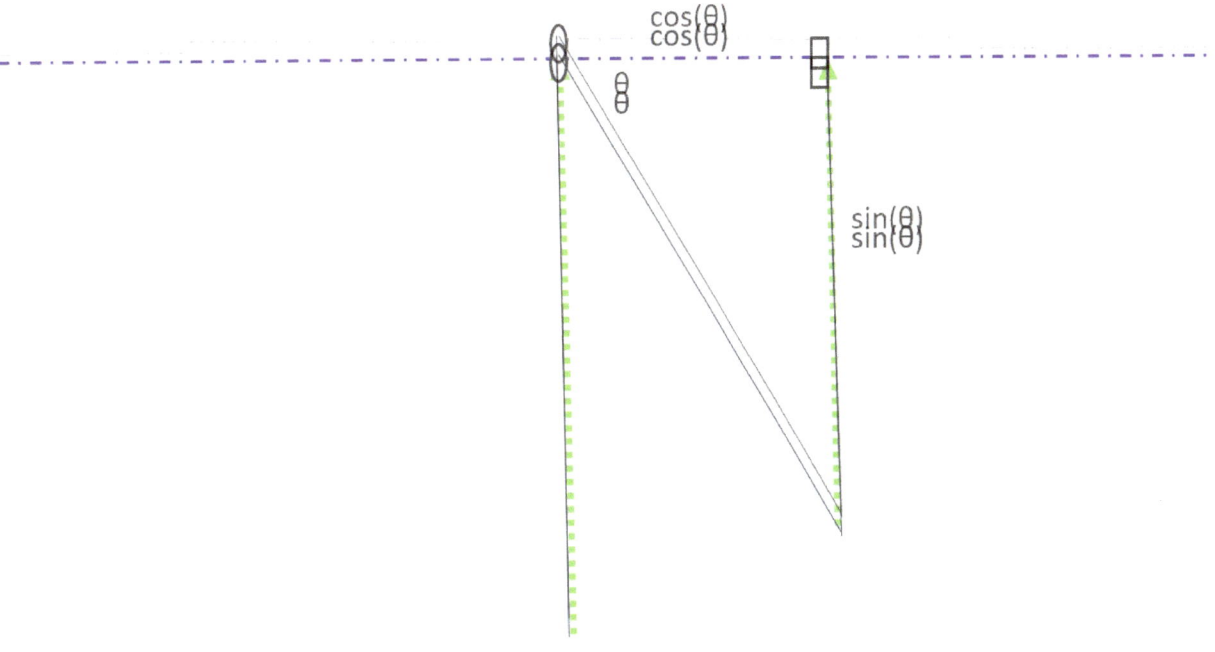

Math Logic for 0.500 and Sine Times Cosine – Beyond Maxwell, Biot-Savart

However, that is not all for the axial nucleostaticmagnetics (or within the close range of an electrical loop as explained by Biot-Savart >200 years ago). The orientation of the direct nucleostaticmagnetics force is repulsive, by its opposition to the attractive proton-electron electrostatic force. However, for the calculation of the axial nucleostaticmagnetics (weak nuclear) force, the rules are more complex for a) inside 10x the electrical loop of Maxwell, Biot-Savart, or for b) inside any atomic system. That is, at any subatomic distance, more 3D engineering is required. A difference in field sources in the other hemisphere becomes important until the 1/distance-squared overwhelms any 1/distance-cubed magnetics.

To analyze, I engineer the force accumulating for the direction is four segments – with the ends in each hemisphere in the opposite direction.

Maxwell found that the rotational force from a magnetic field is based upon the sine of the angle relative to a) the observer object, b) the source loop, c) with the loop's center as the vertex. This will apply later as we get to the rotational force (wave functions) on subatomic particles. The Biot-Savart law, using Maxwell, determine a field strength based upon radius-cubed based upon a movement in a loop of electrons. Both are solid work, and fundamental to understanding here.

However, I will focus, on the less common section from Biot-Savart. It was the earliest version that gets to the 0.500 factor or 2. This factor gets described as 1 at the poles and 2 at the equator. Further, it found a formula that moves from that based upon root(1+3SIN²). The overall field gets described by:

(14)

$$F \equiv \sqrt{1 \mp 3(\sin \theta)^2}^{\frac{1}{2}}$$

It works great. For the linear force (not the rotational to discuss later), the subject here, the linear force is correctly described by that. It goes in strength from 1 to 2 based upon the inclination angle (well, they actually used the longitude angle which is 90 minus the inclination angle. In fact, I will use inclination angle and root(1+3COS²), but those are equivalent by changing from longitude to inclination.

In the most general sense, 0.500 is just the choices where you have a field source that is full or zero (so far away that nothing applies). In that sense, every toward-the-axis source sum will be full half the time.

(15)

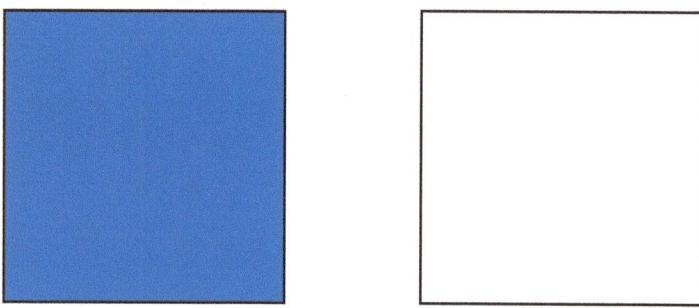

However, that does not help with (x,y,z). So, the evaluation needs sines and cosines; I start with 2D.

I take the sources in four segments because the force of the field in the other hemisphere is different. So, the four segments are:

- The axis sources from the electron (purple sphere) intercept (90-degree drop) to the nucleus (ring) as a blue double arrow
- The same axis distance of sources further (this is a math trick that you will understand soon) as a green double arrow
- The axis sources in the opposite magnetic hemisphere as a red double arrow
- The axis sources in the same hemisphere starting at the same distance right as the red section starts left as a yellow double arrow

(16)

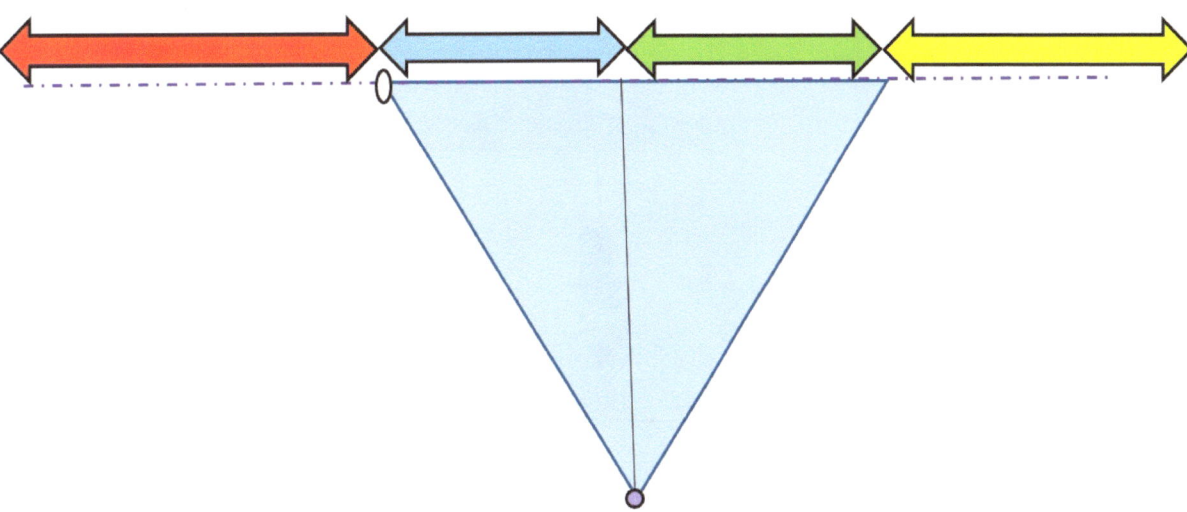

The math trick is to get the red sources and the yellow sources to start at the same distance and both go to infinity. In that sense, those two are exactly the same math, so the calculations will match and aggregate or offset perfectly.

First, the sections of blue and green will offset in the left-right, but aggregate in the upward direction. That is the basic of towards-the-axis strength. That is clean enough, and we could use the classical equations and logic from (14). So, we can ignore the left-right calc, and focus on 2x the calculation from the up towards the axis calculation from just one triangle. (Great, half the work eliminated.)

Now, the real magic of magnetics starts. You see the other hemisphere is different (you know north-south). In the offset of the two additional set of sources, one must think of magnetics. The hemispheres act in opposition. If one is attractive to the poles, the other is repulsive. This is all part of that the field must balance (Gauss, Newton) in the largest sense. While we focus on the near hemisphere for most physics, the far hemisphere impacts the calculation when close to the equator give the 1/distance-cubed balance at the modified Bohr (02-He) radius. (and that, when does magnetics applies = all the time for subatomic physics). This is about the understanding of the nature of subatomic physics as nucleostaticmagnetics and then the time-dependent added forces to that.

So, here is the above logic in vectors. The blue and green arrow offset left-right, but aggregate both towards the axis. By using those segments, the red and yellow arrows offset in both left-right and towards-the axis versus away from the axis. The black arrows show the left-right (white) and versus-the-axis (black) arrow. Please note it remains critical to view all magnetics from the nucleus axis frame-of-reference. The up and down is relative to the y of the electron being positive or negative.

(17)

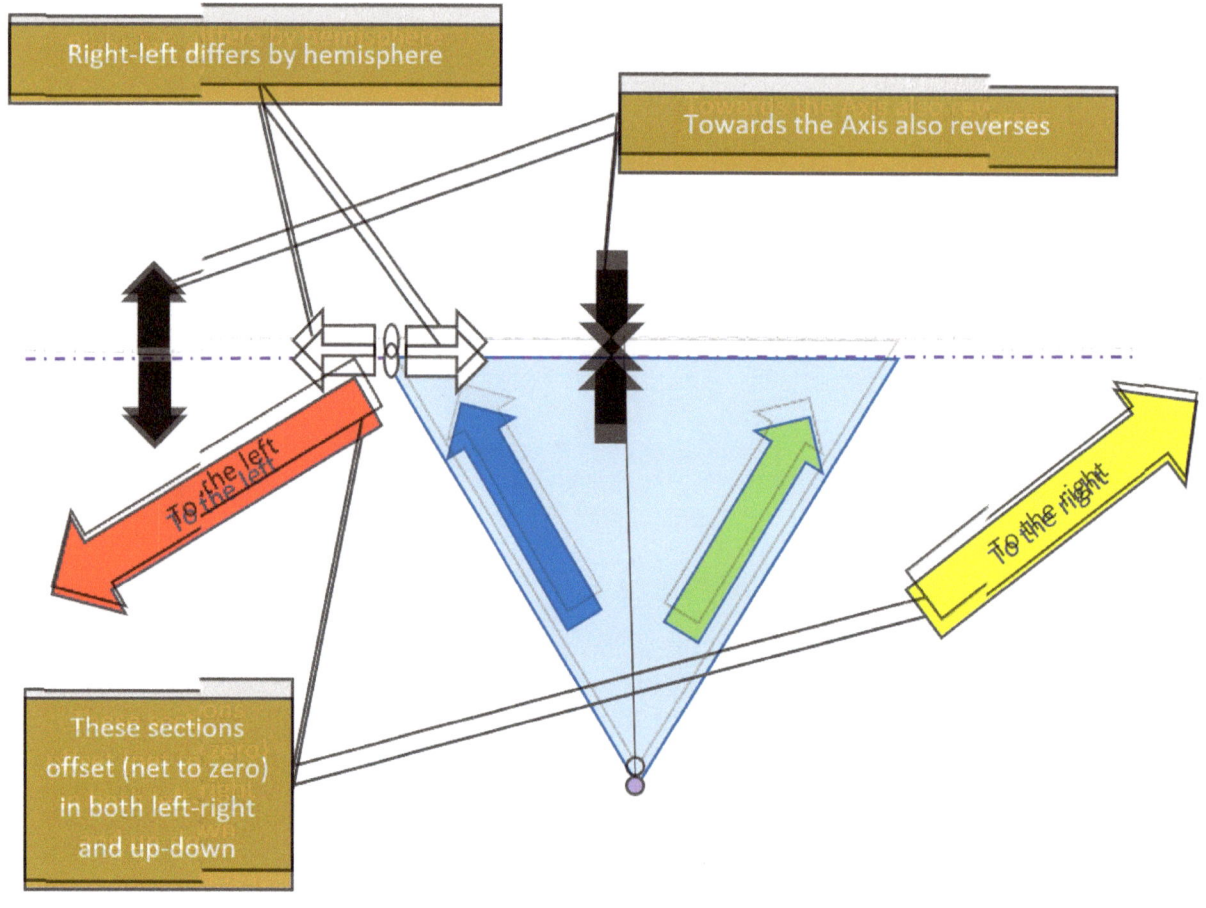

The elimination of the work on red and yellow sections creates focus on the basic triangle towards-the-axis (the blue only, given I can 2x the blue up-down, and ignore the offset left-right). That triangle is the nucleus and particle with the drop and its corresponding as both creating a field in the same hemisphere, and both repulsive towards nucleus (for an electron).

That means that a portion of the force (from the two-hemisphere offsetting segments) need to get excluded from the total force (the magnitude of the axial N-M force, versus the magnitude of the direct N-M force):

That is, the axial N-M force will never be as much as the direct. In fact, it is limited to ½ or less because one loses all the forces down (versus 'up') in the drawing below. However, let's work through the math of that:

This second equation applies to both direct (strong nuclear) and axial (weak nuclear). The direct force is the full 100% calculation above, but the axial force is less by the 3D engineering below: That is, the force is pulled towards the axis at all points, but the ones on the other hemisphere (-) would repulse in that direction (-), so they offset the same range in the closer hemisphere. That leaves a range to 2 * cos(θ) to generate same hemisphere attraction. That creates this 'effective triangle' which is the volume in two equal direction, effectively two triangles out to the particle. Each triangle has the hypotenuse at the distance between particles; it has the axis as one side, and the drop as the 3rd side. The triangle with one side cosine(θ) towards the nucleus matches that same range outward (to the right); are both in the same hemisphere, so contribute (additive). That means the 2 triangles by 2*leg1*leg2 area is the relative strength:

(18)

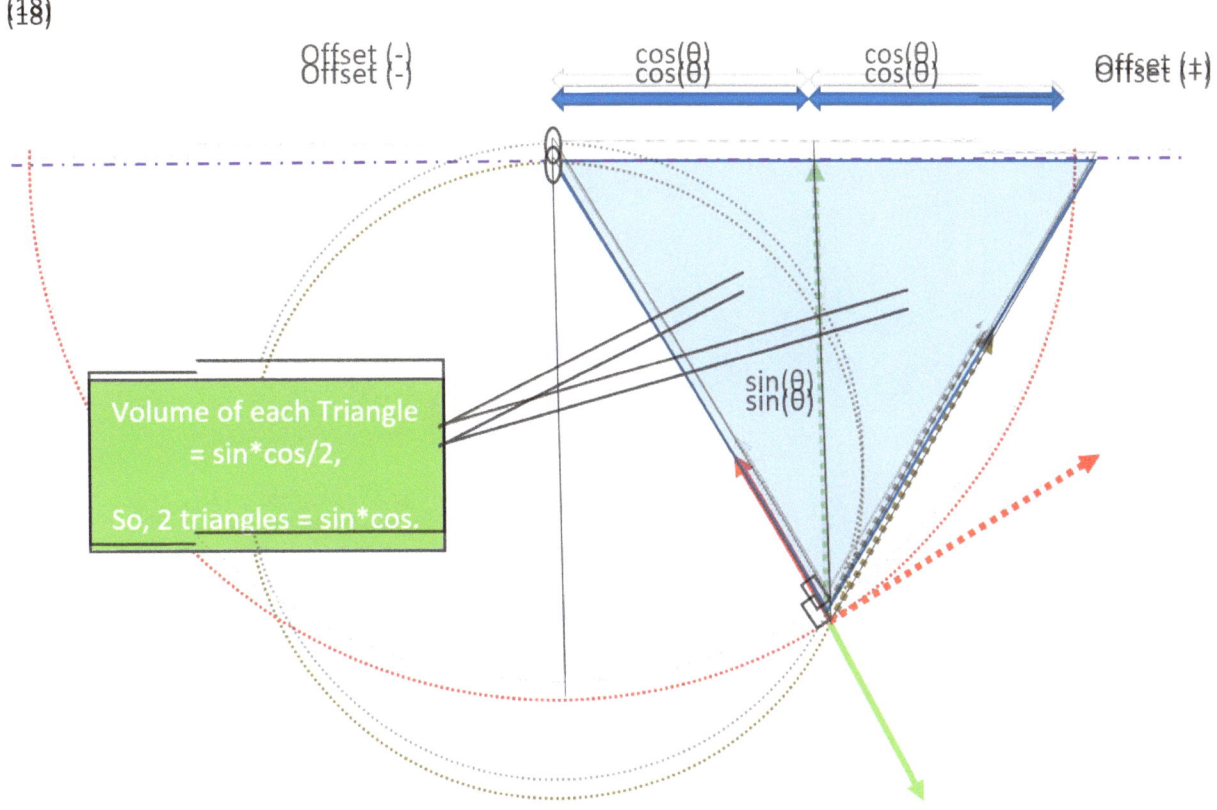

Yet, at the polar position, this triangle is thin and flat, and tiny – by the height. Compare that to (07).

(19)

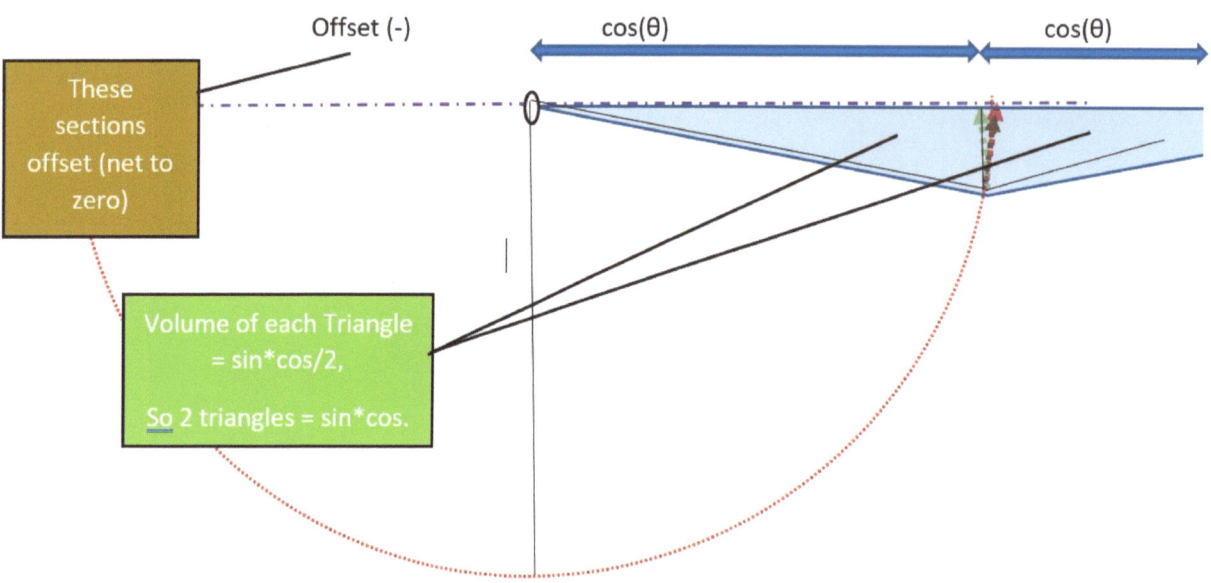

Similarly, near the equator, the triangle becomes relatively small in volume – by the base. Again, smaller than (07).

(20)

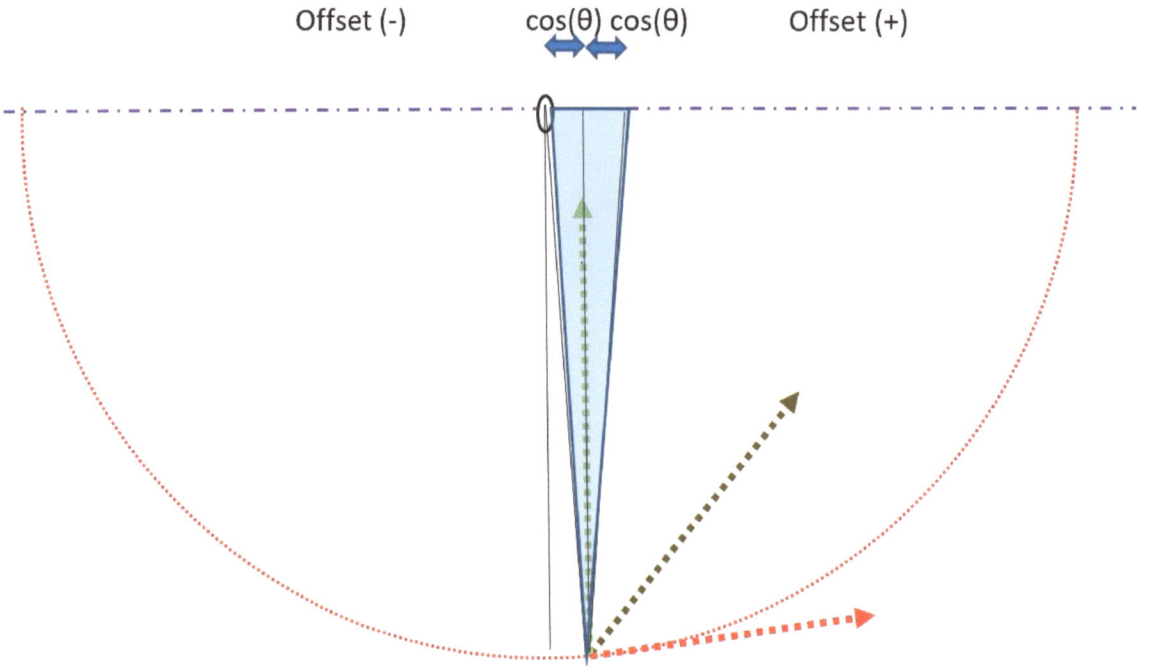

As width is cos(θ), and height is sin(θ), that makes the relative strength sin(θ)*cos(θ).

Notice not just the isotropic (red dotted circle) of electrostatic same-strength field versus the regular toroid (brown dotted circle). The simplification is that the 'average 2-nucelostaticmagnetics' vector (brown hashed arrow) is easier to calculate as if on a toroid (brown dotted circle); it is a tangent.

So, that strength follows the table:

(21)

	xy	z	d	sin(theta)	cos(theta)	Theta (radians)	Theta (degrees)	sin*cos
Equator	1.0000000	-	1.0000000	1.0000000	-	1.5708	90.00	-
	1.0000000	0.0100000	1.0000500	0.9999500	0.0099995	1.5608	89.43	0.0099990
	0.9950000	0.1000000	1.0000125	0.9949876	0.0999988	1.4706	84.26	0.0994975
	0.9800000	0.2000000	1.0002000	0.9798041	0.1999600	1.3695	78.47	0.1959216
	0.9600000	0.3000000	1.0057833	0.9544800	0.2982750	1.2679	72.65	0.2846975
	0.9200000	0.4000000	1.0031949	0.9170701	0.3987261	1.1607	66.50	0.3656598
	0.8700000	0.5000000	1.0034441	0.8670139	0.4982839	1.0492	60.11	0.4320191
	0.7071068	0.7071068	1.0000000	0.7071068	0.7071068	0.7854	45.00	0.5000000
	0.5000000	0.8700000	1.0034441	0.4982839	0.8670139	0.5216	29.89	0.4320191
	0.4000000	0.9200000	1.0031949	0.3987261	0.9170701	0.4101	23.50	0.3656598
	0.5000000	0.8700000	1.0034441	0.4982839	0.8670139	0.5216	29.89	0.4320191
	0.2000000	0.9800000	1.0002000	0.1999600	0.9798041	0.2013	11.53	0.1959216
	0.1000000	0.9950000	1.0000125	0.0999988	0.9949876	0.1002	5.74	0.0994975
	0.0100000	1.0000000	1.0000500	0.0099995	0.9999500	0.0100	0.57	0.0099990
Pole	-	1.0000000	1.0000000	-	1.0000000	-	-	-

Which has a graph of strength of:

(22)

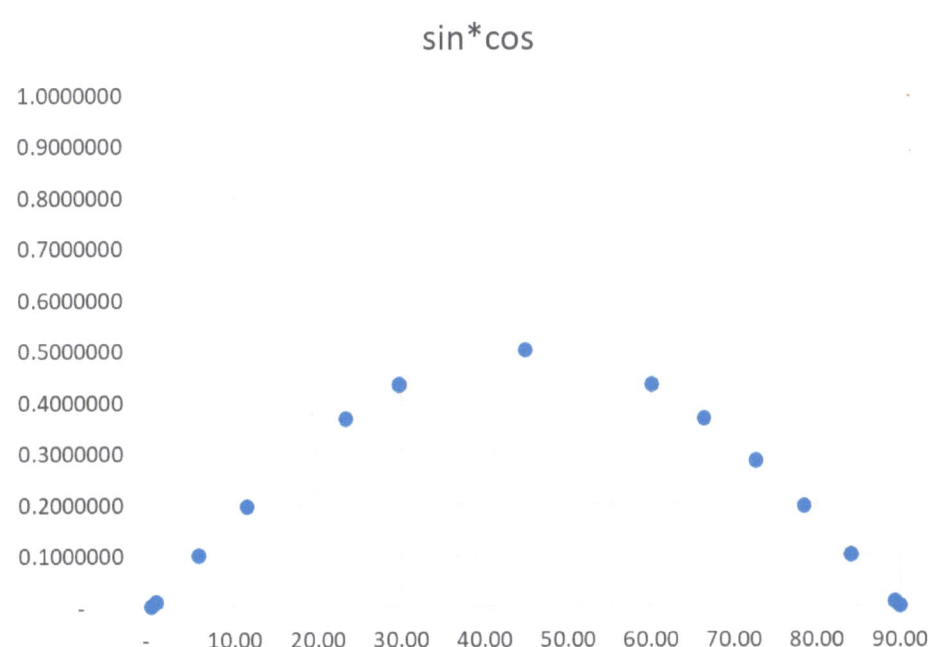

There it is. That 0.500 (½) is the limit of this function which fits the dynamic tension of a subatomic magnetic field. This concept and limit will become very important to the understanding of magnetics.

Postulate Part 1: Within a) 10x the loop radius of a Maxwell-Biot-Savart loop or b) always for all subatomic particle calculations, the calculation of b) traditional magnetic force or b) nucleostaticmagnetics force gets altered from 0.5000 to a sine times cosine factor ($sin(\theta) * cos(\theta)$) (Arno).

So, the **axial nucleostaticmagnetics** force is 'weak' versus the **direct nucleostaticmagnetics** (strong nuclear) force. However, that relative weakness changes by factors not understood before this 3D engineering model. It might be 0.500x or 0.001x, but no matter what that is always the weaker (≤0.5).

- Axial Nucleostaticmagnetics (this postulate – weak nuclear) magnitude at 1/distance-squared ($1/d^3$) based upon magnetic particles (protons, neutrons, and electrons) reduced based upon the segment (triangle) of the axis contributing to the force in the same hemisphere.

(23)

$$\frac{M_1 M_2 * \pm Table(03) * \sin(\theta)\cos(\theta)}{[(x_1-x_2)^2 + (y_1-y_2)^2 + (z_1-z_2)^2]^{\frac{3}{2}}}$$

Which also can get stated with just (x,y,z) as:

(24)

$$\frac{M_1 M_2 * \pm Table(03) * \frac{\sqrt{(x_1-x_2)^2 + (y_1-y_2)^2}}{\sqrt{(x_1-x_2)^2 + (y_1-y_2)^2 + (z_1-z_2)^2}} * \frac{z}{\sqrt{(x_1-x_2)^2 + (y_1-y_2)^2 + (z_1-z_2)^2}}}{[(x_1-x_2)^2 + (y_1-y_2)^2 + (z_1-z_2)^2]^{\frac{3}{2}}}$$

Which then simplifies:

(25)

$$\frac{M_1 M_2 * \pm Table(03) * z * \sqrt{(x_1-x_2)^2 + (y_1-y_2)^2}}{[(x_1-x_2)^2 + (y_1-y_2)^2 + (z_1-z_2)^2]^{\frac{5}{2}}}$$

Of course, the challenge with all subatomic physics is getting a viable observation of that z-direction, the nucleostaticmagnetics axis. Otherwise, the results always end up with a fuzzy ball if the axis is unknown.

Vector Orientation – Axial Nucleostaticmagnetics (Weak) – Why θ/2?

The major improvement postulated here is to create an orientation for each force. That creates a) the physical model, and 3D engineering as the reasons for sines and cosines of wave functions and much more.

That creates the 4-vectors where, for my work, I have chosen the 'z'-direction taken as the frame-of-reference for the **nucleostaticmagnetics axis**, so the combination is easier math as **isotropic** for electrostatic and direct nucleostaticmagnetics (strong nuclear) force, but more complex math for the **towards-the axis** axial nucleostaticmagnetics force. The axial force is always towards z as unchanging (no acceleration in the z-dimension) with the force as (-x) and (-y) to accelerate those back towards the axis. That means the settling positions, or the bottom of any 'pendulum' would be (0,0,z).

However, that gets averaged with the preference for a tangent to the E-S field. So, at the midrange (subshell=p!) the θ/2 force has the most strength.

As noted, each graphic will operate with the z-dimension as the nucleostaticmagnetics. That is true even if I put z= as going to the right to fit it on the page.

The angles are θ from the electron with the nucleus as the vertex, versus the nucleostaticmagnetics axis. Now, since the green vector is 90 degrees, then the other angel is 90-θ (180= θ + 90 + 90-θ with the θ's cancelling). However, the E-S field also operates based upon a 90-degree right angle, so that means that the angle between the hashed vectors is back to θ (90-θ + θ = 90).

(26)

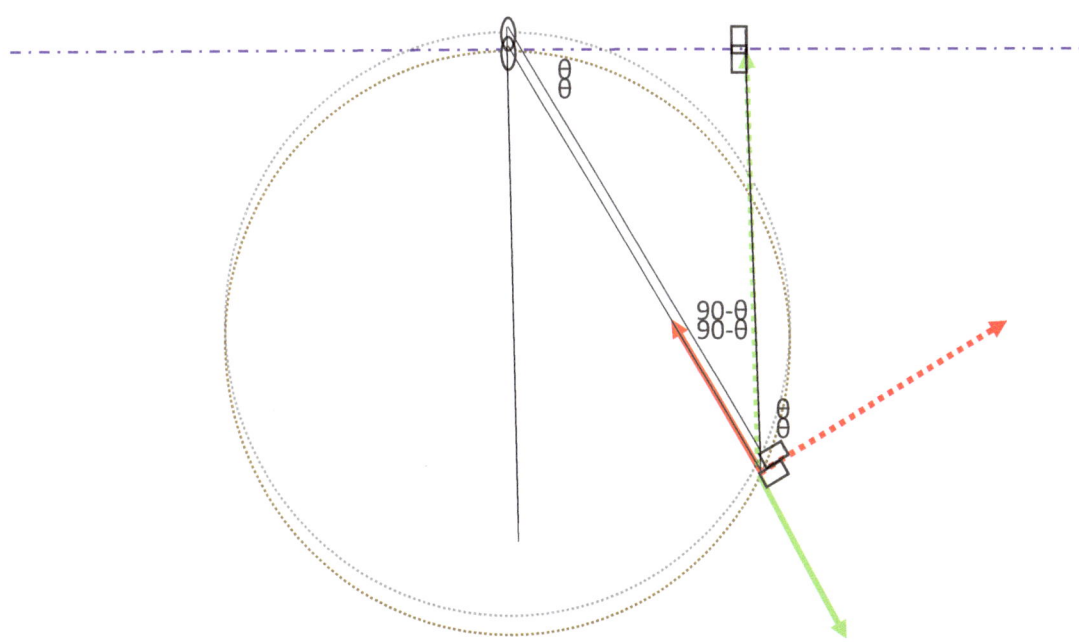

However, these are both applied, so one would add the vectors. That makes an equal sides parallelogram, and that means that the angles to the final, combined forces is ½ of that angle above.

As such, one see θ/2 as the angle for wave functions in quantum mechanics.

(27)

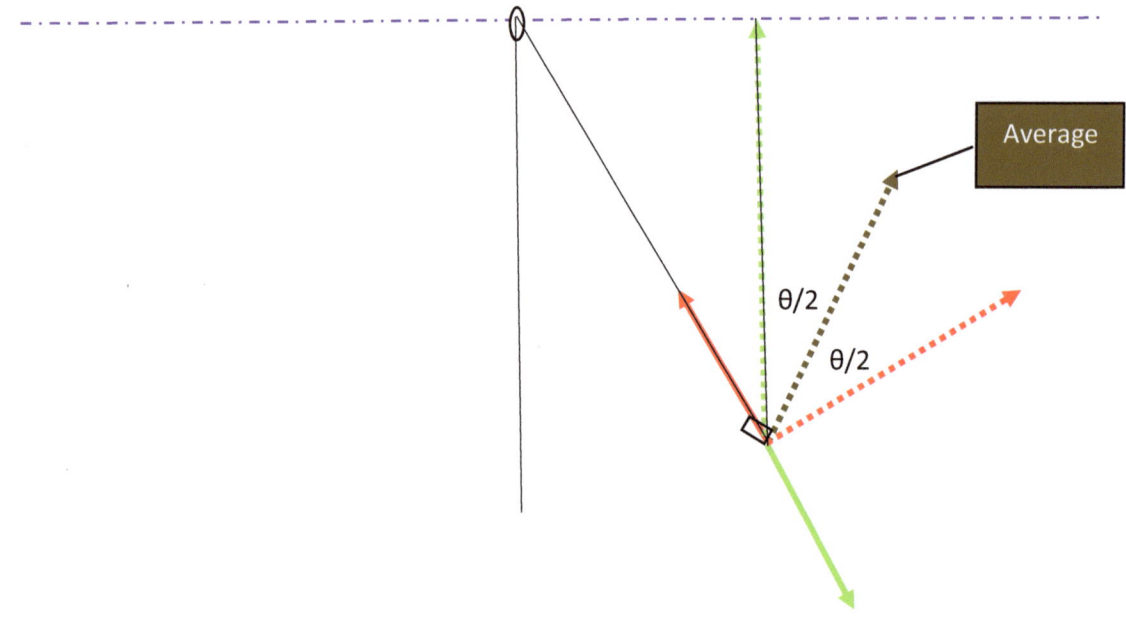

The electron will try to use both vectors (overlay then find the same strength on that aggregation), and that means it operates in a direction at θ/2 of the axis perpendicular (the drop to the axis). In effect, the fields overlay (superposition), and the path is the average.

Postulate Part 2 – For the 3D direction of fields, one must superposition all the applicable fields to get the preferred path that follows the preference of aggregation for the same strength field direction. Once adding magnetics (inside 10x or for any subatomic event), this is not the 90-degree orthogonal applicable in traditional method used for macro-electrostatic events.[v]

For electrostatic fields alone, that is easy. The same strength is the tangent plane – any direction at 90 degrees to the particle-particle direction.

However, for nucleostaticmagnetics (and traditional magnetics), this overall force is towards-the-axis which might be an any angle. Mostly, I think of it as an energy valley going towards the nucleostaticmagnetics axis. Going to left or right (in the above drawings) would get extra sideways forces, and higher forces overall.

This combination concept explains the preferred direction, although one must keep remembering that way scientists called 'weak nuclear' force is really a combination of two different types of force (really fields for direction). The E-S vector is 90 degrees from the particle-particle direction, but the the N-M vector is towards the nucleostaticmagnetics axis. Those are not the same, with the on-the-axis exception, of course; and will cause many challenges as we work through first movement, then fields, then magnetism.

So, I review the inclination angles for those linear vectors for the three (4) forces (I save the rotational forces for later) for the nucleus-electron interaction.

- Direct electrostatic is isotropic physics-negative, so the direction is (-Δx,-Δy, -Δz).

- Direct nucleostaticmagnetics is a direct opposite, so the direction is also (Δx, Δy, Δz).

- Axial is at 90 degrees (π/2 radians) to the axis, so (-Δx, -Δy, 0).

- Electrostatic field is at 90 degrees (π/2 radians) to the direct electrostatic. This is the famous dot-product math. However, that plane gets focused on the nucleostaticmagnetics intersection, so it is that dot-product where x=0 and y=0, so on the z-axis. That math is a little easier.

So, there is not just the magnitude, and some balancing of those.[vi] That is, because the field is both direct and axial, its field for the net-strength is ½ between the two. As such, the aggregation is at θ/2 and not θ. The field overlays the axis orientation and the axis+electron angle and averages!

(28)

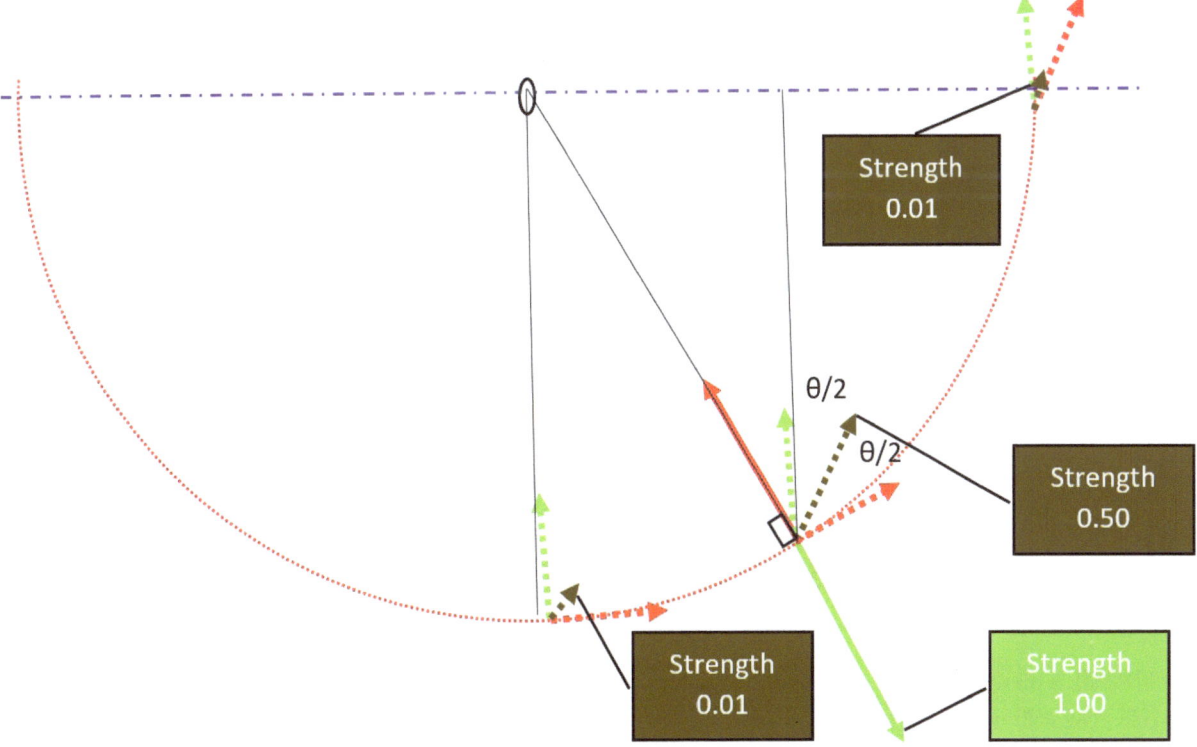

Postulate Part 3 – Electrons want to build a) at the poles or b) at the equator. The poles (and Pauli-hemisphere pairs) are sets of two, one in each hemisphere. The equatorial settling positions are up to three, so usually in odd-count Elements, and only transitional, not found in full subshell configurations (like most even-count Elements such as Column 8, Column 2). That creates the band gap added spacing that lowers electrical resistance, and thereby increase electrical conductivity (for example Column 11).

Please Note: For matrix algebra, the θ/2 toroid 'cheat' makes the calculation easier. The direction of the average is as if on a toroid, but it also is confusing as the force do not continue inward after 90 degrees, like the toroid. The toroid does not center on the nucleus except >45. As such, I prefer my 4-vector model to avoid that engineering error.[vii]

(29)

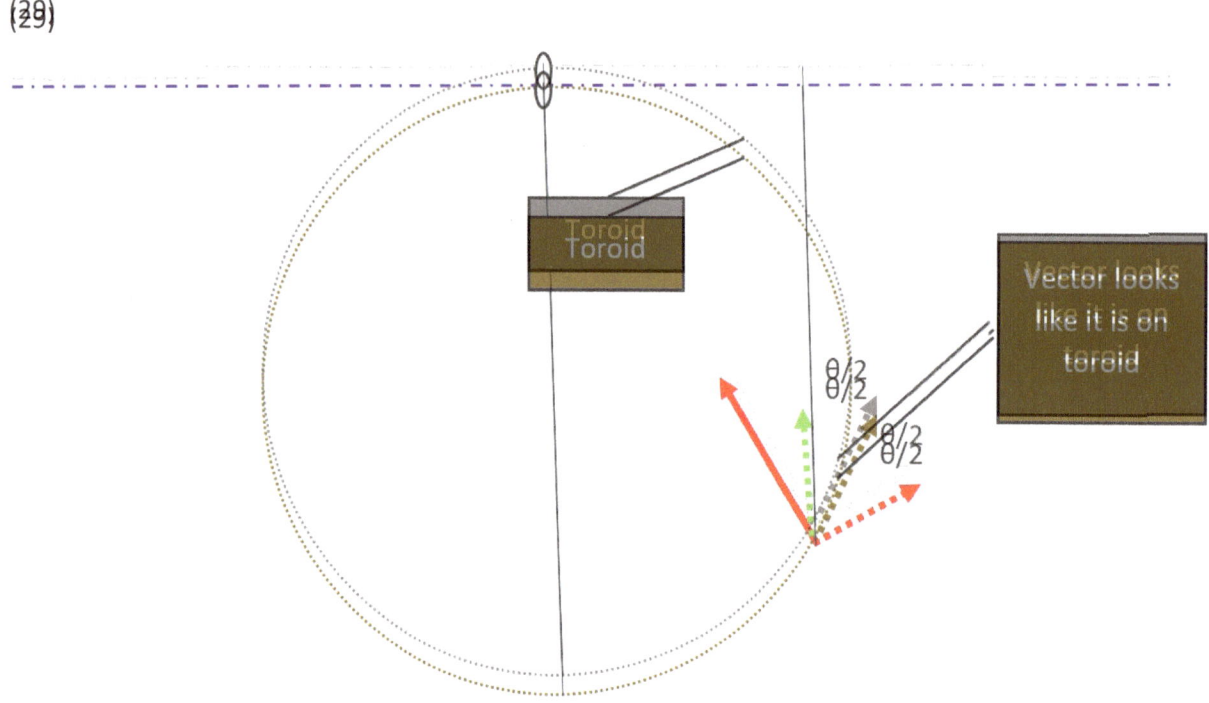

That toroid will come back into the work in the time-dependent model, but it does apply to the static model here in the static model at all. It is θ/2, but not a toroid.

This creates a position for the vector that is towards the pole, but also at the spherical balance of electrostatic and direct nucleostaticmagnetics (strong nuclear) forces. The interesting thing is that the direction (vector) is actually inside that as the particle position gets closer to the equator:

Let's review. Near the poles, that vector is definitely pointed close to the ES-Direct NM field equilibrium. Yet, because of the θ/2 as one gets to the poles, the vector is pointed to inside the ES-Direct NM field equilibrium.

That creates two distances to get the axis intercept of the axial N-M vector. First, one has the cosine(θ) as the distance out for the electron positions. However, the further distances is based upon sine(θ/2) for the averaging[viii] ES + Direct NM. Further, that 2nd distance gets based upon the distance off the axis which is sine(θ). That makes segment-2 as sine(θ)* sine(θ/2) (column 5 = column 3 * column 4). These are added as the SUM-Direction on Axis (column 6 = column 2 + column 5). Finally, the vector Strength remains reduced by that sine(θ)*cosine(θ) factor (column 2 * column 3):

(30)

Theta	cos(theta)	sin(theta)	sin(theta/2)	sin(theta/2)*sin(theta)	SUM-Direction	Strength
-	1.000000				1.000000	0.0%
0.5	0.999962	0.008727	0.004363	0.000038	1.000000	0.9%
1.0	0.999848	0.017452	0.008727	0.000152	1.000000	1.7%
2.0	0.999391	0.034899	0.017452	0.000609	1.000000	3.5%
3.0	0.998630	0.052336	0.026177	0.001370	1.000000	5.2%
4.0	0.997564	0.069756	0.034899	0.002434	0.999999	7.0%
5.0	0.996195	0.087156	0.043619	0.003802	0.999996	8.7%
10.0	0.984808	0.173648	0.087156	0.015134	0.999942	17.1%
15.0	0.965926	0.258819	0.130526	0.033783	0.999708	25.0%
20.0	0.939693	0.342020	0.173648	0.059391	0.999084	32.1%
25.0	0.906308	0.422618	0.216440	0.091471	0.997779	38.3%
30.0	0.866025	0.500000	0.258819	0.129410	0.995435	43.3%
35.0	0.819152	0.573576	0.300706	0.172478	0.991630	47.0%
40.0	0.766044	0.642788	0.342020	0.219846	0.985891	49.2%
45.0	0.707107	0.707107	0.382683	0.270598	0.977705	50.0%
50.0	0.642788	0.766044	0.422618	0.323744	0.966532	49.2%
55.0	0.573576	0.819152	0.461749	0.378242	0.951819	47.0%
60.0	0.500000	0.866025	0.500000	0.433013	0.933013	43.3%
65.0	0.422618	0.906308	0.537300	0.486959	0.909577	38.3%
70.0	0.342020	0.939693	0.573576	0.538986	0.881006	32.1%
75.0	0.258819	0.965926	0.608761	0.588018	0.846837	25.0%
80.0	0.173648	0.984808	0.642788	0.633022	0.806670	17.1%
85.0	0.087156	0.996195	0.675590	0.673019	0.760175	8.7%
86.0	0.069756	0.997564	0.681998	0.680337	0.750094	7.0%
87.0	0.052336	0.998630	0.688355	0.687411	0.739747	5.2%
88.0	0.034899	0.999391	0.694658	0.694235	0.729135	3.5%

Here is it in graphic. The closer to the equator, the more the axial force vector (brown hashed arrows) drifts inward from the equilibrium pole.

(31)

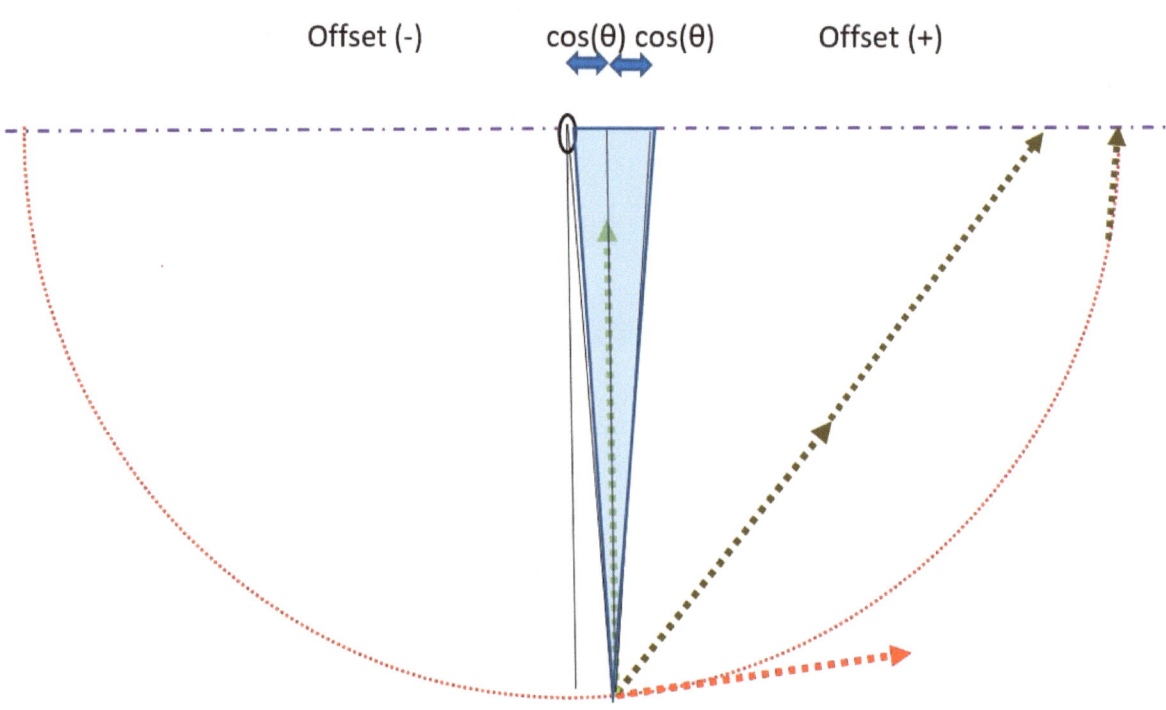

As the angle increase towards the equator, the intercept drifts inward from the . That is a solid reason why subshell-d (I call it subshell-u) and subshell-f (my -v) are inside the polar first pair (-s which I call -m for magnetic axis). The x-direction in angle; the y-direction is ratio inside equilibrium,

(32)

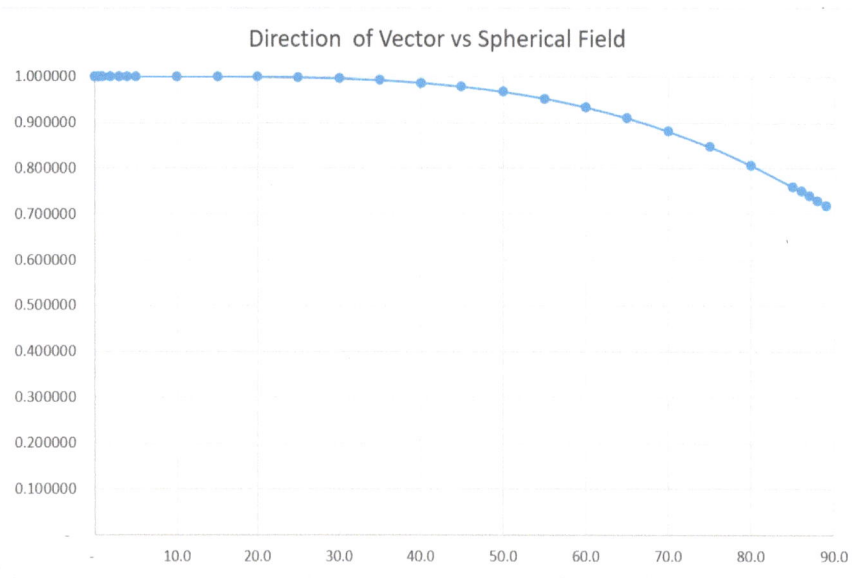

Think about the impact on electrons settling positions (subshells). The inclination angles would generally get distributed from pole to equator to opposite hemisphere pole. By the equilibrium of electrostatic (E-S) and nucleostaticmagnetics (N-M strong nuclear), that would be the same distance, so all the same energy. That is, even spacing, even nucleus distance, same subshell energy

(33)

Old Naming (electrons as same inclination)	AVSC N-M Subshell Naming	Inclination Angle
6s1	6m1	0.00
6p1,2,3	6t1-3	25.71
5d01,02,03,04,05	6u1-5	51.43
4d01,02,03,04,05,06,07	6v01-07	77.14
4d8,9,10,11,12,13,14	6v08-14	102.86 (77.14)
5d06,07,08,09,10	6u06-10	128.57 (51.43)
6p4,5,6	6t4-6	154.29 (25.71)
6s2	6m2	180 (0.00)

However, this weak force vector creates a system where the settling positions operate by the 4-vectors, so the electrons are not all at the same inclination, yes, in subshells at defined inclination angles, but also the electrons gets pushed by that direction of the 2-field-vectors.

This is a little complex.

1. First, the strength of the vector is different at different inclination.
2. Second, the direction changes as the sine(θ/2); that means it is a little closer at the outside. However, there is a nuance in the subshell-p as that vector actually goes closer, then ends further out.
3. Finally, especially, with the subshell-p, the electron-electron force in 3D of the other subshells also push the final positions slightly.

Please note that I have a different reasoning, thereby disagree with the Aufbau strict filling order where the shell numbers of subshell-d and subshell-f are in layers below. Those are closer, yes, but the -d and -f subshells are in the same master subshell. They only have the closer distance because of the axial part of the 4-vectors.

So, for subshell 6, the 6s electrons settle at the poles, and settle at the electrostatic + direct nucleostaticmagnetics equilibrium. I rename that as subshell-m for on the **m**agnetic poles as **6m**.

Then, the 6p electrons settle at the poles, and settle at the electrostatic + direct nucleostaticmagnetics equilibrium. I rename that as subshell-t as a **t**ight endcap (small inclination angle) as **6t**.

So, the 5s electrons settle at the poles, and settle at the electrostatic + direct nucleostaticmagnetics equilibrium. I rename that as subshell-u for the **u**pper inclination angle and thereby as **6u**.

So, the 6s electrons settle at the poles, and settle at the electrostatic + direct nucleostaticmagnetics equilibrium. I rename that as subshell-v for the **v**ery large inclination angle and thereby as **6v**.

(34)

Old Naming (electrons as same inclination)	AVSC N-M Naming	Inclination Angle	Distance
6s1	6m1	0.00	Equilibrium
6p1,2,3	6t1-3	25.71 minus a little	Equilibrium + as per #2, #3
5d01,02,03,04,05	6u1-5	51.43 minus a little	Equilibrium – largest vector
4d01,02,03,04,05,06,07	6v01-07	77.14 minus a little	Equilibrium – smaller vector
4d08,9,10,11,12,13,14	6v08-14	102.86 (77.14)	Equilibrium – smaller vector
5d06,07,08,09,10	6u06-10	128.57 (51.43)	Equilibrium – largest vector
6p4,5,6	6t4-6	154.29 (25.71)	Equilibrium + as per #2, #3
6s2	6m2	180 (0.00)	Equilibrium

So, here is the net vectors. The subshell-6t (-p) is slightly out, and the 6u and 6v quite inward. It has two steps, the basic axial force (brown hashed arrow), plus the E-S repulsions (red hashed arrows) of surrounding electrons.

Note that the one closer to the equator are basically about the same the E-S direction (1/2 of a very small angle). However, near the equator, the θ/2 drives the electron in closer. That makes subshell-d (-u) and subshell-f (-v) inside subshell-s (-m). Those are shown on the left hemisphere example.

However, there is also E-S repulsion from other electrons. That is shown on the right side. So, the first off-axis subshell-p (-t for tight endcap) gets pushed along the E-S field by the axial N-M, but then combination those slightly inward -d (-u) and -f (-v) electrons, plus the direct off axis of the -s (=m) create a subshell-p which is beyond subshell-s (-m)

(35)

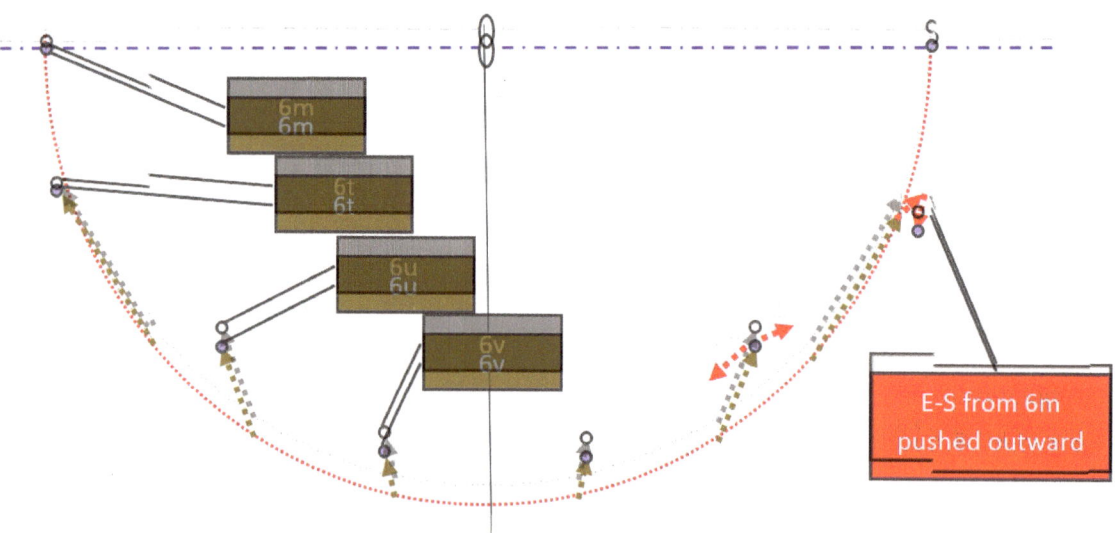

Note that the calculation of subshell settling positions is a) E-S and Direct N-M equilibrium, plus b) other subshell's electron positions. Please remember this above is 2D, and those electrons have different latitudes, so the forces need another sine factor for strength. However, the 6m (-s) on the axis is always 100% outward, so again that 6p outward makes sense. Note that for 6p, the force pushed that electron outside the equilibrium (red dotted sphere). With the above combination of vectors, the observation of =d and =f being inside is correct.

Vector for Each Force

The vector for electrostatic force for Particle-2 is (please note that (x_1-x_2) is the same as –(x_2-x_1) which some papers and textbooks prefer) combines the **magnitude** with the **normalized direction**. For electrostatic forces, that is reasonably easy since the magnitude is a constant times the product of the number of charge particles in each objects (particle-set).

(36)

$$F = \begin{bmatrix} \dfrac{(x_2 - x_1) * kQ_1Q_2}{(x_2 - x_1)^2 + (y_2 - y_1)^2 + (z_2 - z_1)^2} \\ \dfrac{(y_2 - y_1) * kQ_1Q_2}{(x_2 - x_1)^2 + (y_2 - y_1)^2 + (z_2 - z_1)^2} \\ \dfrac{(z_2 - z_1) * kQ_1Q_2}{(x_2 - x_1)^2 + (y_2 - y_1)^2 + (z_2 - z_1)^2} \end{bmatrix}$$

The vector for direct nucleostaticmagnetics (strong nuclear) force for Particle-2 is very similar to Coulomb with the change in that a) the denominator at the **cubed** factor and b) the numerator calc based upon **nucleostaticmagnetics (protons + neutrons in the nucleus) particles**:

(37)

$$F = \begin{bmatrix} \dfrac{(x_2 - x_1) * M_1M_2}{((x_2 - x_1)^2 + (y_2 - y_1)^2 + (z_2 - z_1)^2)^{\frac{3}{2}}} \\ \dfrac{(y_2 - y_1) * M_1M_2}{((x_2 - x_1)^2 + (y_2 - y_1)^2 + (z_2 - z_1)^2)^{\frac{3}{2}}} \\ \dfrac{(z_2 - z_1) * M_1M_2}{((x_2 - x_1)^2 + (y_2 - y_1)^2 + (z_2 - z_1)^2)^{\frac{3}{2}}} \end{bmatrix}$$

Remember that nucleostaticmagnetics particles include neutrons, so the basis counts are different for the nucleus side. Further, this becomes the fundamental that will build into a) mass, b) magnetism, and other combination attributes and forces.

However, the 3rd vector with the double requirement: a) towards the axis (the drop), plus b) orthogonal, which creates the θ/2. We need to calculate the extra distance along the axis for the intercept. What is the '?' distance below given we have θ$_{NM}$.

Because of the two fields, the direction gets calculated from 2θ$_{NM}$.

Again, I will proceed with the static model of the electron in different positions around the nucleus. That is complex enough for now.

The challenge is to generate the orientation of the axial nucleostaticmagnetics force. Specifically two challenges. First, the force is towards the closest axis, and that flips when the electron crosses the equator, a huge discontinuity in the math to address. Second, the force direction wants to keep that 2θ net-2-field relationship stable. It wants the path of least change. So, that means the angle is not a simple 90 degree (direct towards the axis), but a more complex average net-2-field calculation.

To calculate that, I will calculate the z-direction and the xy-direction for that regular toroid by some good old fashion geometry proof methods. I will calculate the orientation based upon the particle position and the other particle-set's (nucleus) nucleostaticmagnetics axis. Everything works in the nucleus frame-of-reference. We have enough challenges without adding the whole set rotational that is gas state (the average used in Hartree) or other molecules or fields.

However, that now gives a calculation for the triangle with '?' in it. Because of the x,y drop at 90 degrees (π/2 radians) to both parallel z-axis lines, the angle of the vector towards the axis versus towards the nucleus is 90-θ. And, since the E-S field is orthogonal (90-degress), the two that we average go back to θ [90-(90-θ)].

However, those are relative the same, so the average, the path of least change for movement is generally θ/2.[ix]

Further, since we have the angle, and we know the off-axis distance as root(x^2+y^2), we can calculate the '?' and the hypotenuse to generate the normalized vector of the axial N-M.

(38)

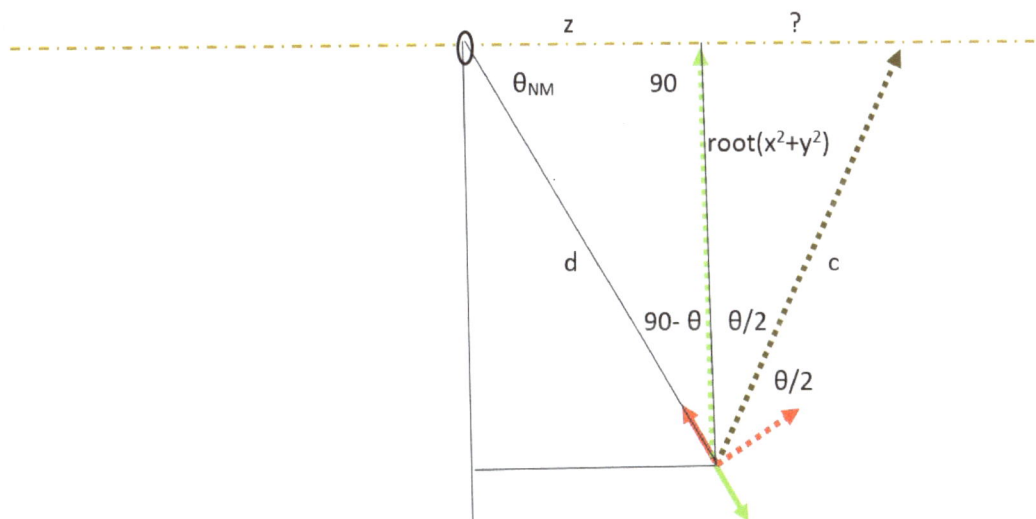

Now, the angle at 'a' must be 180-θ/2. Since the angle at 'b' is 90, and the total must equal 180, then at the particle position, that angle is θ/2 from the right angles (90=2*(θ/2)+90-θ).

The 'z' direction '?' for vector direction (the normalized triangle) would calculate by calculating the sides for the (root,?,c) triangle which both must have the angle (θ/2), so we use the tangent functions for find '?' since we know the root(x,y) distance:

(39)
$$\tan\left(\frac{\theta}{2}\right) \equiv \frac{?}{\sqrt{x^2 + y^2}} \quad \therefore \quad ? \equiv \frac{\tan\left(\frac{\theta}{2}\right)}{\sqrt{x^2 + y^2}}$$

So, the hypotenuse (c) would be the Pythagorean Theorem applied to those two factors: * and **

*with a hemisphere factor to add later
**with a rotation factor to add after that

(40)
$$c \equiv \sqrt{x^2 + y^2 + \frac{\left(\tan\left(\frac{\theta}{2}\right)\right)^2}{x^2 + y^2}}$$

So, that makes the **normalized** vector, the direction without magnitude:

(41)
$$\begin{bmatrix} \dfrac{-x}{\sqrt{x^2 + y^2 + \frac{\left(\tan\left(\frac{\theta}{2}\right)\right)^2}{x^2 + y^2}}} \\[2em] \dfrac{-y}{\sqrt{x^2 + y^2 + \frac{\left(\tan\left(\frac{\theta}{2}\right)\right)^2}{x^2 + y^2}}} \\[2em] \dfrac{\frac{\tan\left(\frac{\theta}{2}\right)}{\sqrt{x^2 + y^2}}}{\sqrt{x^2 + y^2 + \frac{\left(\tan\left(\frac{\theta}{2}\right)\right)^2}{x^2 + y^2}}} \end{bmatrix}$$

Notice that the x and y do not change as all of these calculation are locked into that x,y versus the z-axis plane. All these operations (until we get to the time-depending traditional magnetics) are in that plane.

That makes the vector for the axial N-M (weak) force, the normalized * the magnitude:

So, the combination of the normalized vector with the magnitude gets the axial N-M vector. The z-dimension seems perfect with that tangent function with the above revision. The tangent with the θ/2 = 90 gets the jumps (discontinuities) and the zeros correctly placed.

So, the full vectors are the normalized vector times the magnitude as follows:

(42)

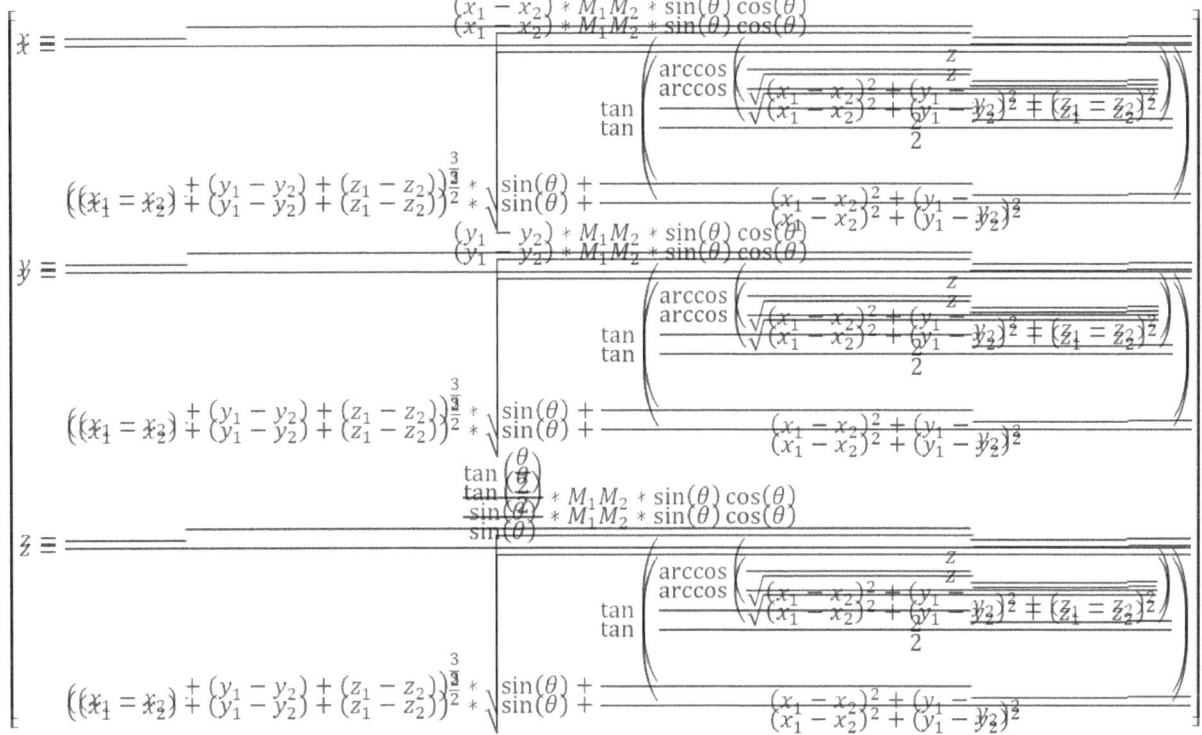

Or if trying to use computational methods, which is my work, one must expand these based upon x,y,z).

(43)

$$\begin{bmatrix} x = \dfrac{(x_1 - x_2) * M_1 M_2 * \sin(\theta)\cos(\theta)}{((x_1-x_2)+(y_1-y_2)+(z_1-z_2))^{\frac{3}{2}} * \sqrt{\sin(\theta) + \dfrac{\tan\left(\dfrac{\arccos\left(\dfrac{z}{\sqrt{(x_1-x_2)^2+(y_1-y_2)^2+(z_1-z_2)^2}}\right)}{2}\right)}{(x_1-x_2)^2+(y_1-y_2)^2}}} \\[2em] y = \dfrac{(y_1 - y_2) * M_1 M_2 * \sin(\theta)\cos(\theta)}{((x_1-x_2)+(y_1-y_2)+(z_1-z_2))^{\frac{3}{2}} * \sqrt{\sin(\theta) + \dfrac{\tan\left(\dfrac{\arccos\left(\dfrac{z}{\sqrt{(x_1-x_2)^2+(y_1-y_2)^2+(z_1-z_2)^2}}\right)}{2}\right)}{(x_1-x_2)^2+(y_1-y_2)^2}}} \\[2em] z = \dfrac{\dfrac{\tan\left(\dfrac{\theta}{2}\right)}{\sin(\theta)} * M_1 M_2 * \sin(\theta)\cos(\theta)}{((x_1-x_2)+(y_1-y_2)+(z_1-z_2))^{\frac{3}{2}} * \sqrt{\sin(\theta) + \dfrac{\tan\left(\dfrac{\arccos\left(\dfrac{z}{\sqrt{(x_1-x_2)^2+(y_1-y_2)^2+(z_1-z_2)^2}}\right)}{2}\right)}{(x_1-x_2)^2+(y_1-y_2)^2}}} \end{bmatrix}$$

The interesting math is that for all three (x,y,z) dimensions, the denominator has that tangent function. It is squared, then a square root, but effectively that means that the it become infinity, so the forces gets small by a division by tangent (2θ-90) at the same orientations as the z-dimension infinities.

However, at both the infinite positions of the tangent function it multiplies by either sine or cosine which is zero. As such, it should be immaterial.

Does an infinity over infinity offset? Does an infinity time zero equal zero? Well, let's see.

If so, that makes the major flip as less material, even immaterial. Here is an example calculations for 'z' getting close to 90 degrees for θ and thereby 180 degrees for 2θ

I focus this section of the analysis on a 2D slice of the above. This is the toroid circle which then is at any (x,y) combination. The (x,y) generates a distance which, with the inclination angle, defines the toroid. The z-direction I take as the nucleostaticmagnetics (strong/weak nuclear) force axis.

The above vector is much more complex than a simple tangent.

- The tangent in the denominator drives the vector calculation to a tiny number, especially at the orientations of large tangents (0, 90, 180, and so on)
- The 'z' axis component has the tangent in both the numerators and denominator. That is strange.

Further, the force for x,y at the poles becomes immaterial. Here is example calculations for 'z':

(44)

x,y = root(x^2+y^2)	z	d	sin(theta)	cos(theta)	tan(theta)	-xy	z	c	norm(-xy)	nomr(z)
0.010000	1.000000	1.000050	0.010000	0.999950	0.010000	(0.010000)	(0.000050)	0.010000	(0.99998750)	(0.00499931)
0.020000	1.000000	1.000200	0.019996	0.999800	0.020000	(0.020000)	(0.000200)	0.020001	(0.99995005)	(0.00999450)
0.030000	1.000000	1.000450	0.029987	0.999550	0.030000	(0.030000)	(0.000449)	0.030003	(0.99988777)	(0.01498146)
0.704000	0.707000	0.997730	0.705602	0.708609	0.995757	(0.704000)	(0.205749)	0.733450	(0.95984760)	(0.28052200)
0.705000	0.707000	0.998436	0.706105	0.708108	0.997171	(0.705000)	(0.205950)	0.734466	(0.95988108)	(0.28040740)
0.706000	0.707000	0.999142	0.706606	0.707607	0.998586	(0.706000)	(0.206150)	0.735482	(0.95991472)	(0.28029221)
0.707107	0.707107	1.000000	0.707107	0.707107	1.000000	(0.707107)	(0.206349)	0.736600	(0.95996007)	(0.28013688)
0.708000	0.707000	1.000556	0.707606	0.706607	1.001414	(0.708000)	(0.206548)	0.737513	(0.95998247)	(0.28006008)
0.709000	0.707000	1.001264	0.708105	0.706107	1.002829	(0.709000)	(0.206746)	0.738529	(0.96001658)	(0.27994315)
0.710000	0.707000	1.001973	0.708602	0.705608	1.004243	(0.710000)	(0.206943)	0.739544	(0.96005083)	(0.27982565)
1.000000	0.030000	1.000450	0.999550	0.029987	33.333333	(1.000000)	(0.021175)	1.000224	(0.99977588)	(0.02117033)
1.000000	0.020000	1.000200	0.999800	0.019996	50.000000	(1.000000)	(0.014131)	1.000100	(0.99990017)	(0.01412942)
1.000000	0.010000	1.000050	0.999950	0.010000	100.000000	(1.000000)	(0.007070)	1.000025	(0.99997501)	(0.00706948)
1.000000	0.001000	1.000000	1.000000	0.001000	1,000.000000	(1.000000)	(0.000707)	1.000000	(0.99999975)	(0.00070711)
1.000000	(0.001000)	1.000000	1.000000	(0.001000)	(1,000.000000)	(1.000000)	(0.000707)	1.000000	(0.99999975)	(0.00070711)
1.000000	(0.010000)	1.000050	0.999950	(0.010000)	(100.000000)	(1.000000)	(0.007070)	1.000025	(0.99997501)	(0.00706948)
1.000000	(0.020000)	1.000200	0.999800	(0.019996)	(50.000000)	(1.000000)	(0.014131)	1.000100	(0.99990017)	(0.01412942)
1.000000	(0.030000)	1.000450	0.999550	(0.029987)	(33.333333)	(1.000000)	(0.021175)	1.000224	(0.99977588)	(0.02117033)
			#DIV/0!	#DIV/0!						
(0.030000)	(1.000000)	1.000450	(0.029987)	(0.999550)	0.030000	0.030000	0.000449	0.030003	0.99988777	0.01498146
(0.020000)	(1.000000)	1.000200	(0.019996)	(0.999800)	0.020000	0.020000	0.000200	0.020001	0.99995005	0.00999450
(0.010000)	(1.000000)	1.000050	(0.010000)	(0.999950)	0.010000	0.010000	0.000050	0.010000	0.99998750	0.00499931
(0.001000)	(1.000000)	1.000000	(0.001000)	(1.000000)	0.001000	0.001000	0.000000	0.001000	0.99999988	0.00050000
(0.000100)	(1.000000)	1.000000	(0.000100)	(1.000000)	0.000100	0.000100	0.000000	0.000100	1.00000000	0.00005000
0.000100	(1.000000)	1.000000	0.000100	(1.000000)	(0.000100)	(0.000100)	(0.000000)	0.000100	(1.00000000)	(0.00005000)
0.001000	(1.000000)	1.000000	0.001000	(1.000000)	(0.001000)	(0.001000)	(0.000000)	0.001000	(0.99999988)	(0.00050000)
0.010000	(1.000000)	1.000050	0.010000	(0.999950)	(0.010000)	(0.010000)	(0.000050)	0.010000	(0.99998750)	(0.00499931) Repeat to top

Let's review this in three sections.

At the 45-degree position 0.707107 for (x,y) and z (upper green section). At that point the (x,y) points back towards the axis, and the 'z' axis is zero. There is a steady movement from outward and towards the extreme right on z (nucleostaticmagnetics axis), though that straight towards the axis, then towards the extreme left on z (nucleostaticmagnetics axis).

At the 90-degree position where (x,y) is near 1 and z is near zero (second green section), the transition from positive to negative is not a discontinuity in (x,y) because of the large calc of the denominator.

At the 180-degree position where (x,y) is near -1 and z crosses zero, one notices the discontinuity of the normalized angle for (x,y) jumping from +1.00 to -1.00 (shown in red). However, the strength of the force (orangish) is tiny, so that discontinuity occurs at the time when the force has no impact. As such, there might not be much physical impact. Further, the axis (z-) direction is quite stable. However, that gets masked because in most cases, we are examining the normalized, so that is always 1, so it can look huge.

The normalized looks like a discontinuity, but it has not physical 3D engineering impact. (Ah, all of quantum methods are normalization, so they get discontinuities. This resolves the quantum dilemma.)

In this way, the tangent does not create infinity discontinuities.

This occurs because that tangent really is based upon the original θ at the same time, and I can simplify all that potential of infinity from tangent so just sines and cosine which are not discontinuous.

That changes the denominator at (38) which we worried about as going to zero; it is impossible for the denominator to achieve that zero because it is either off one direction (x,y) or off the other direction (z being the N-M axis) even with the θ/2.

(45)
$$\sqrt{x^2 + y^2 + \left(\frac{\tan(\theta)}{\tan(\frac{\theta}{2})}\right)^2}, so\ worried\ x^2 + y^2 + \left(\frac{\tan(\theta)}{\tan(\frac{\theta}{2})}\right)^2 \equiv 0$$

I analyze this in two parts. The first is a sine function, and the 2nd is a square, so never negative. The first part is 0 only at the equator angle (90, 270). So, the only locations for potential concern would be 45, 135, and such. Those being 45 = 90/2.

So, what is the tangent of 45 degrees. That is 1 and -1 which both square to 1.

As a result, the denominator is always based upon 0+1 or something that is not zero plus a square.

Note that because the (03) rule of a duopole, the closest axis is always 90 degress or less. It flips by (03) once it gets past the equator. I do not need to examine the 135 degree position as the (03) flipped that back to 45 relative the closest axis.

No Electrons Falling into the Nucleus

That the function does not go through infinity discontinuities is one great part.

However, I also noticed that with those factors, the axial N-M force is never directly back to the nucleus; that means that the axial N-M never offsets the direct N-M. The 'electrons never fall into the nucleus'; that equilibrium is modified a) for off axis positions, and b) for multiple electron-pairs sytems, but the failure of the Bohr model is completely avoided.

So, for the theoretical physicists, not computational mathematicians like me, I will state as the normalized function. That the first part of that (44) equation is really sin(θ) also, so the denominator normalized can get analyzed as a 2-dimensional matrix:

(46)
$$\begin{bmatrix} (x,y) \equiv \dfrac{\sin(\theta)}{\sqrt{\sin(\theta) \mp \dfrac{\tan\left(\frac{\theta}{2}\right)}{\sin(\theta)}}} \\ z \equiv \dfrac{\tan\left(\frac{\theta}{2}\right)}{\sqrt{\sin(\theta) \mp \dfrac{\tan\left(\frac{\theta}{2}\right)}{\sin(\theta)}}} \end{bmatrix}$$

And the vector with strength becomes:

(47)
$$\begin{bmatrix} (x,y) \equiv \dfrac{\sin(\theta) * M_1 M_2 * \sin(\theta)\cos(\theta)}{(d)^{\frac{3}{2}} * \sqrt{\sin(\theta) \mp \dfrac{\tan\left(\frac{\theta}{2}\right)}{\sin(\theta)}}} \\ z \equiv \dfrac{\tan\left(\frac{\theta}{2}\right) * M_1 M_2 * \sin(\theta)\cos(\theta)}{(d)^{\frac{3}{2}} * \sqrt{\sin(\theta) \mp \dfrac{\tan\left(\frac{\theta}{2}\right)}{\sin(\theta)}}} \end{bmatrix}$$

Again, I think the direction of 3D engineering path is better, but I expect that theoretical physicists will need to contemplate this in the abstract math starting from the above.

This causes strange behavior of electrons within the subatomic distances, and the 3D engineering atomic model that I explain later. The full understanding of electron forces, inside at atomic distances, are not direct and isotropic (from the particle in straight lines to the other particles). Further, they are not all 1/distance-squared so easily analyzed with matrix addition.

I recognized forces that change dramatically moving electrons a) towards the poles first, and b) in strange paths in interactions, and c) in helical patterns once settled into a settling position in large sets. However, these strange and amazing paths get based upon 3D engineering.

It is a system that have math integrity and sound 3D engineering.

Analyzing Rotating Electron in Stable Settling Position Alone Still Gets Tangent

Please note that the above is electrons in different static positions; it is not the dynamic time-dependent model. The time-dependent first adds an electron rotating. That is where the complexity occurs. You see, that is strength of repulsion depending on the inclination angle – if all factors constant (the nucleus) and rotate the duopole electron, and **one electron rotating still gets a tangent function**.

(48)

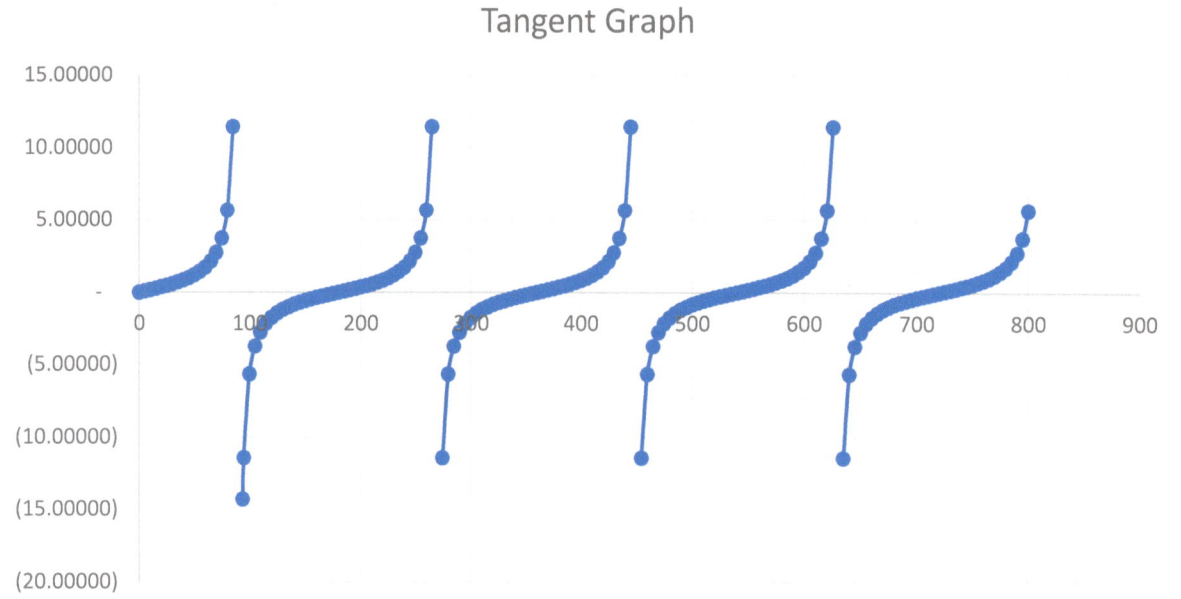

The above graph is fine for a) only axial N-M force, and b) only for the rotation of the electron. It has a limited purpose. That is, this is just axial nucleostaticmagnetics (weak nuclear) force for a rotating electron. It excludes the other force (crossed-out) vectors.

(49)

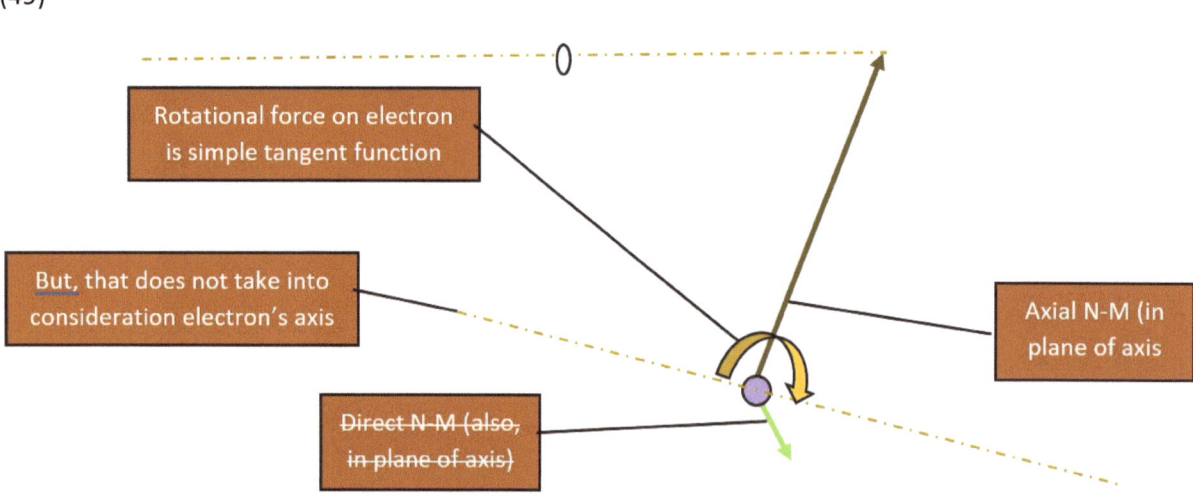

The above (remember the confusion where you used Dirac's tangent of $\theta_{toroid}/2$ – which is also in my drawings as θ_{N-M}) is the orientation of the closest axial force on the electron as it rotates. In that way, it fits the basic requirements. However, that does not consider a) direct nucleostaticmagnetics (strong nuclear) force, or b) any impact of the electron's only axis as it rotates. These are the more complex issues that this postulate will address step-by-step.

Further, and more important, the simple tangent model does not take into account that the electron (subatomic particle) itself has two poles. It is, by the above field equations, a **duopole**. As such, the tangent function has a repetition that happens in a 2x cycle (that being noted by both Dirac and Bose and elsewhere throughout quantum mechanics as those 'well, . . . just calculate' physics procedures).

Dirac's $\theta/2$ is the repeating nature from my electron duopole concept, but that also happens to be the magnetic duopole stabilizing field equation. In my work, **while both equations have the same $\theta/2$, they are not the same equation**. In application, especially when one gets to the equator quantum leap to the other pole as closest, those equation take two different tracks. The rotational equation does not reverse signs, but the axial (towards-the-pole) equation does. At that point, we get many of the 'just calculate' challenges solved by the postulated 3D physical model.

Dirac's tangent of $\theta_{toroid}/2$ is correct for the simple force of the electron rotating (the basic wave function), but the above system has four (4) interlocking forces. The axial nucleostaticmagnetics (weak nuclear) force provides both a linear force on the distant particle-set and a rotational force for the particle-set at the vertex. Then the axis of the 2nd particle-set provides those two forces in reverse. I introduce these 4 concepts, but I focus on the electron linear force here, and will handle rotational in more detail later.

However, the $\theta/2$ equation takes strange discontinuities for electron movement over an equator. To explain, the axial force generates four forces all the time. Two forces on the two particles for two axis orientations. The complexity is that each particle has a) linear force towards the other axis and b) a rotational force to rotate its own axis to align with the other particle positions (not to align with axis – *with exception). The linear force is an <u>orange arrow</u>. The rotational force is a <u>black curved arrow</u>.

(50)

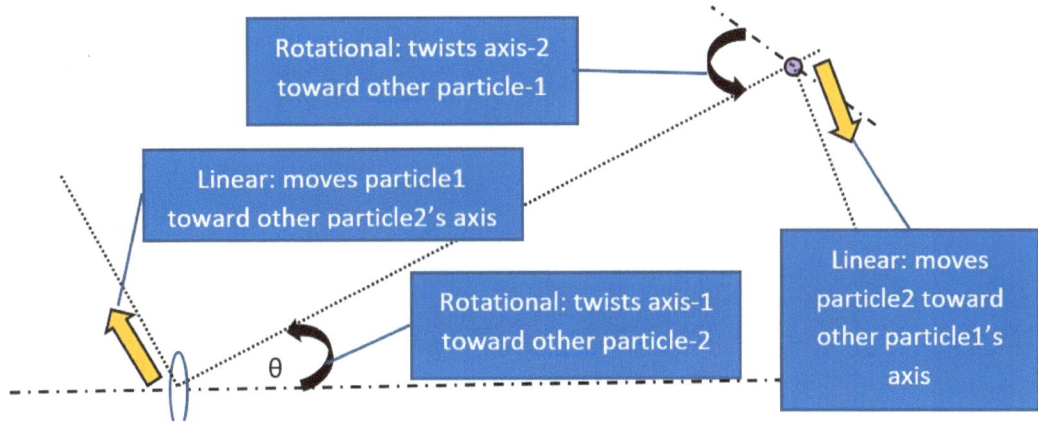

However, back at the tangent, the 3D engineering has negative infinity at 0, and the cycle is 180 above, where the duopole must repeat over a 90 degree cycle. To make a complete picture, I need a vector with the linear forces (I skip rotational force for the moment, but keep a marker for later), not just for the positions, but also for direction. That is the leap from an abstract number, to a 3D engineering vector. So, using the angle calculated from (09), the graph present both a) the associated angles for discontinuities, and b) the associated sign of the slope for the z-direction properly.

The duopole gets to the behavior that Dirac mistakenly called a monopole, but one might make that guess from the above view. Dirac was correct in the phenomenon, but wrong in the physical concept = a duopole instead of a monopole.

(51)

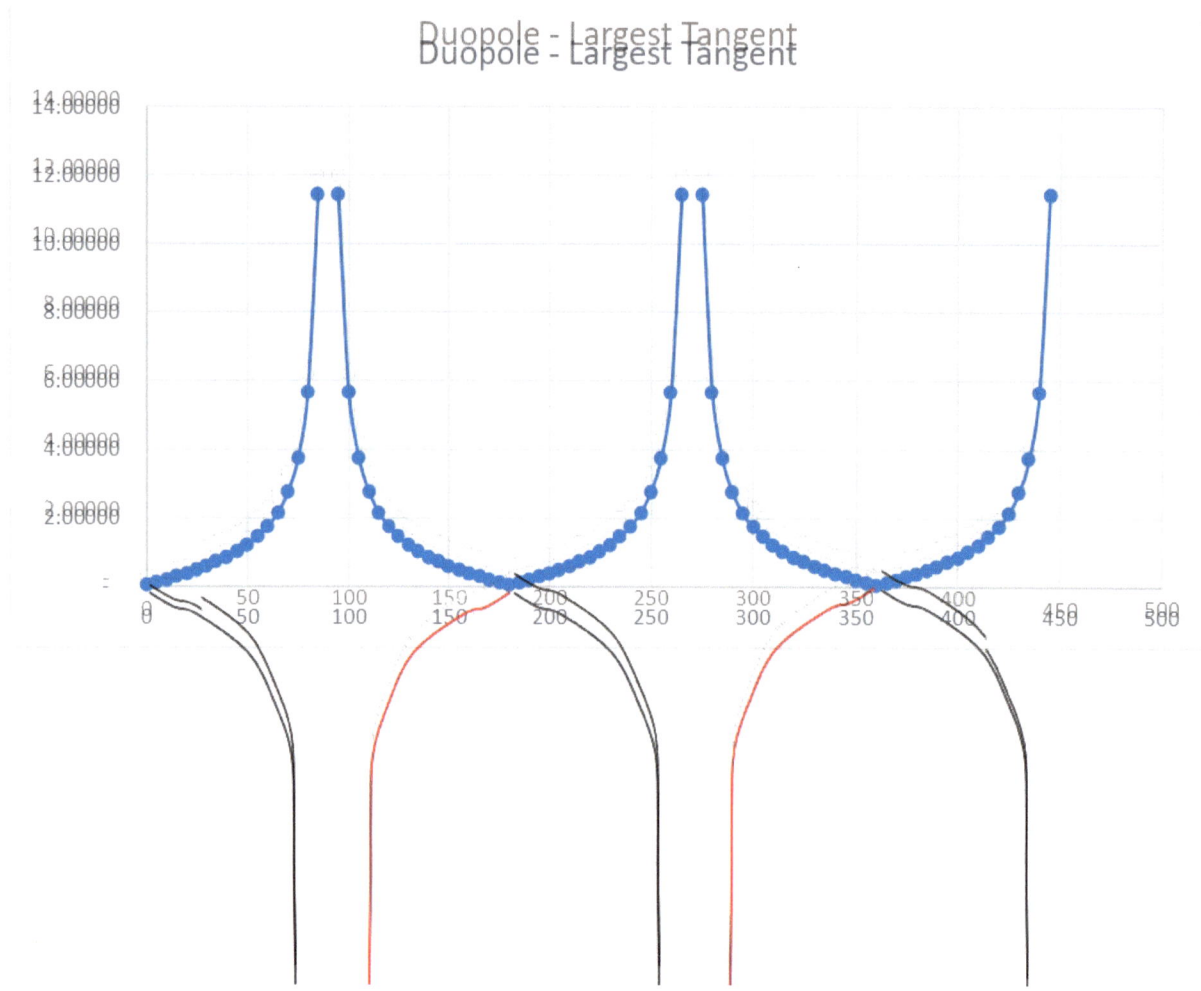

However, this analysis was focused until the angle got to 90, and the 2θ got to 180. However, there is a challenge with that. The nearest axis is a huge leap at the equator.

The cycle of forces in math would flip every other cycle (the θ only). And we end with a mismatch in future cycles. There is a mismatch, not in magnitude, but in the positive versus negative sign.

Final Thought on Dirac

Yet, the Dirac use of φ was incomplete. That is the orientation of the drop or really about the sphere at different latitudes at the same longitude. It really is the changes along a chosen latitude where θ is the inclination/longitude angle. However, Dirac was only thinking as abstract degrees of freedom. φ is a 2nd degree of freedom, but depending on the atomic system, its direction is not orthogonal because of the combination = which Dirac assumed. In part, he stated this by allowing the more complex. With the above equations, we have a physical model to generate the relationship of Dirac's 4th, still valid, with its 3D engineering in a physical model of layers, inclination/longitude, latitude, and hemispheres.

Further, that the electron with its two poles would be a tangent, but that is not the '3D pendulum' but the very small movements at θ/2.

It might be understood by my separate paper in a clean transition from Quantum Numbers into 3D hemispherical coordinates.

(52)

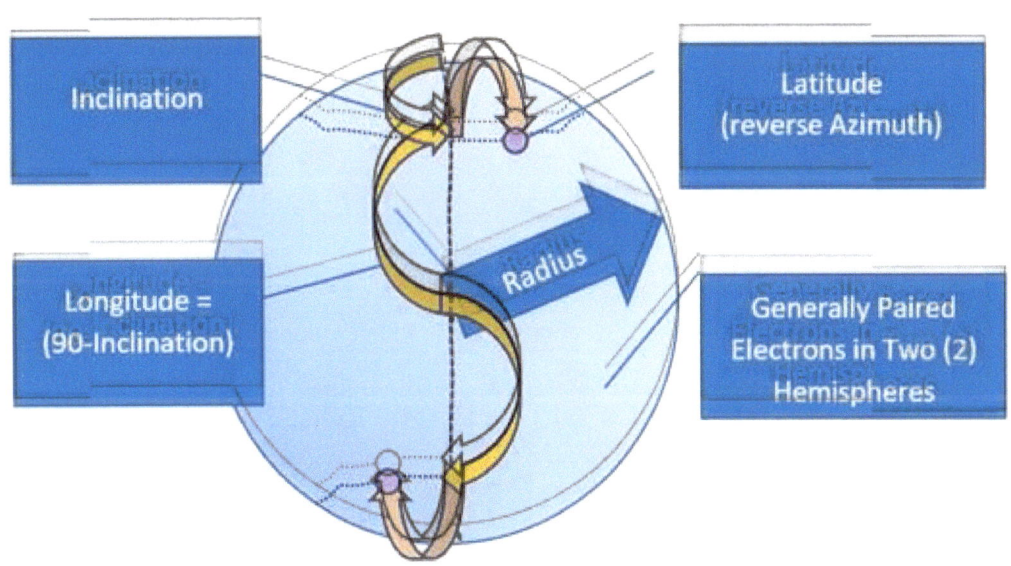

Physics Positive Versus Negative Forces

Now, the work is not quite complete. The signs remain a challenge. I worked with absolute values. Let me explain with a picture of the quadrants.

(53)

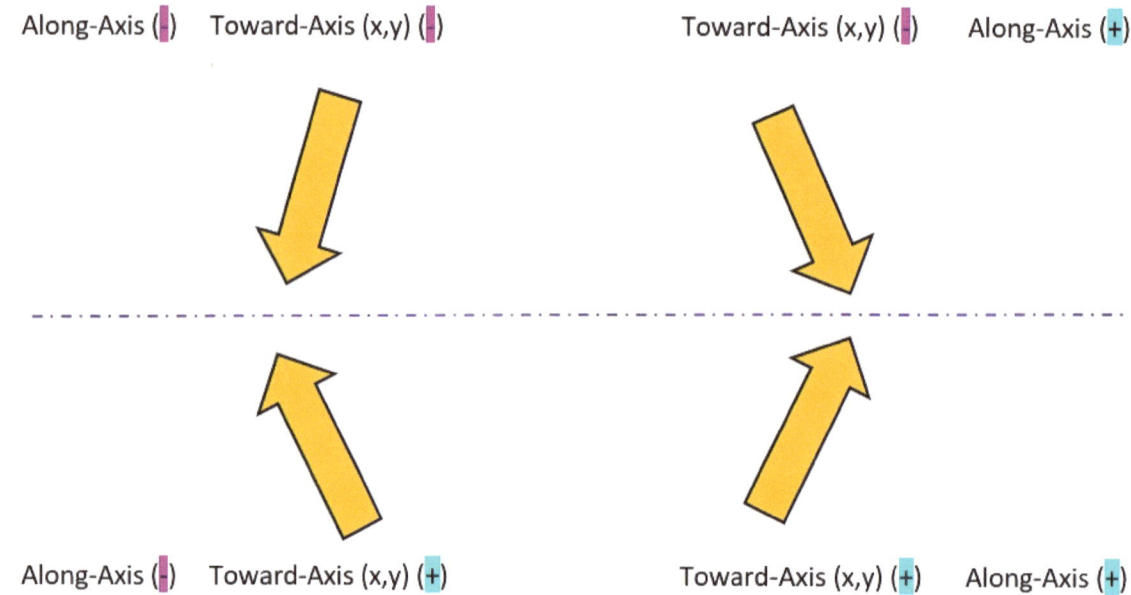

So, what happens in common terms, around a clock.

(54)

(xy,z) on a clock	Toward Axis (2θ)	Along the Axis (z-)
(+0.1,+1.0) after 12	Negative (-)	Positive (+)
(+1.0,+0.1) just shy of 3	Negative (-)	Positive (+)
(+1.0,-0.1) after 3	Positive (+)	Positive (+)
(+0.1,-1.0) just shy of 6	Positive (+)	Positive (+)
(-0.1,-1.0) just after 6	Positive (+)	Negative (-)
(-1.0,-0.1) near 9	Positive (+)	Negative (-)
(-1,+0.1.0) just after 9	Negative (-)	Negative (-)
(-0.1,+1.0) near 12	Negative (-)	Negative (-)

The orientation 'quantum leap' towards the opposite axis once the angle goes past the equator (the 12 and 6 on the clock). But, that is when force is tiny (sin*cos). The same with the change from towards-the-axis change as 3 and 9, but again that is when the force is tiny (sin*cos).

So, the N-M weak nuclear force vector is the vector product of the normalized direction with the magnitude.

(55)

$$F = \begin{bmatrix} \dfrac{(x_1 - x_2) * M_1 M_1}{((x_1 - x_2)^2 + (y_1 - y_2)^2 + (z_1 - z_2))^{\frac{3}{2}} \sqrt{x^2 + y^2 + \dfrac{\left(\tan\left(2\theta - \frac{\pi}{2}\right)\right)^2}{x^2 + y^2}}} \\ \dfrac{(y_1 - y_2) * M_1 M_2}{((x_1 - x_2)^2 + (y_1 - y_2)^2 + (z_1 - z_2))^{\frac{3}{2}} \sqrt{x^2 + y^2 + \dfrac{\left(\tan\left(2\theta - \frac{\pi}{2}\right)\right)^2}{x^2 + y^2}}} \\ \dfrac{\dfrac{\tan\left(2\theta - \frac{\pi}{2}\right)}{\sqrt{x^2 + y^2}} * M_1 M_2}{((x_1 - x_2)^2 + (y_1 - y_2)^2 + (z_1 - z_2))^{\frac{3}{2}} \sqrt{x^2 + y^2 + \dfrac{\left(\tan\left(2\theta - \frac{\pi}{2}\right)\right)^2}{x^2 + y^2}}} \end{bmatrix}$$

That drives a 3D engineering atomic model building from the poles (subshell-s of up to two electrons only) as the tightest 3D balance of the 3-force equation. There is a settling position at each pole, and there is a net-2-axial-force 'weak nuclear' force pushing them into those position.

(56)

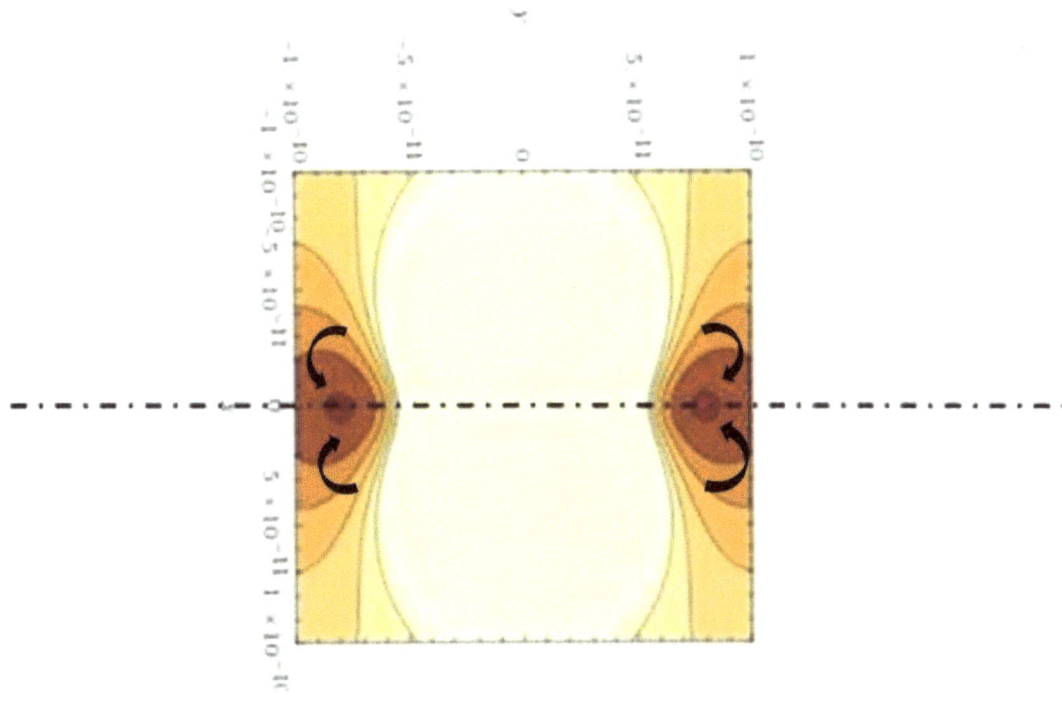

The above is a solution to the Schrödinger Equation, which is not the 'fuzzy ball', but generates similar results (if one does not have a knowable z-dimension). Notice that the electrons do not fall into the nucleus, but has a density away from the nucleus.

If you rotate this randomly, the results are the less dense electrons field versus the proton field. (That is the way the quantum theory solves for the lack of a weak nuclear force for electrons and directional vectors.) This solution tracks with quantum theory, but comes from a 3D engineering physical model.

Which leads to inclination angles for subshells building in two hemispheres to explain the Periodic Table of Elements. After the first two electrons, the subshells build from equator to poles at the same inclination angle, and that generates different chemical properties, bonding angles, and everything about chemistry and chemical engineering.

(57)

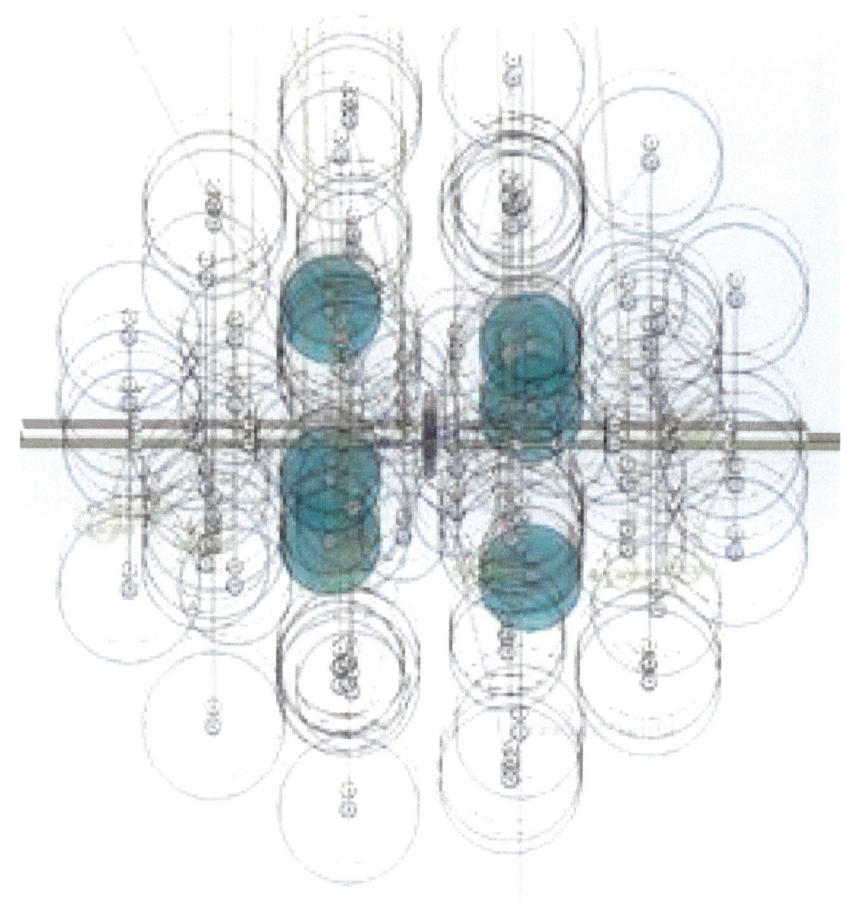

Of course, then further applications get time-dependent models that will also work.

So, the math nature of nucleostaticmagnetics is a constant torque on the particles to move towards the axis, and to move in this counterclockwise (ah, right-hand) helical (Compton) small wave functions depending on the electron position and velocity. However, the above has the other electrons with electrostatic repulsions creating the walls of an energy well that locks particles into settling positions.

Finally, this gives a continuous 3D engineering model for molecular bonding and energy wells (negative valance):

The outer electron of one atom finds an open pathway to the nucleus of a 2nd atom. It locks into that position by the electron-proton electrostatic (opposites attract) force (blue arrow). However, it cannot get too close by the N-M repulsion, so there is a equilibrium at the modified Bohr radius (green double arrow). However, the other outer electrons of both atoms have electron-electron electrostatic (like-kind repel) repulsion (red arrow). So, the bonding position is the 3D engineering of these force:

Of course, these surrounding electrons provide sideways force (yellow arrow) for 3D stability so make bonding angles very consistent for the same Elements combining.

(58)

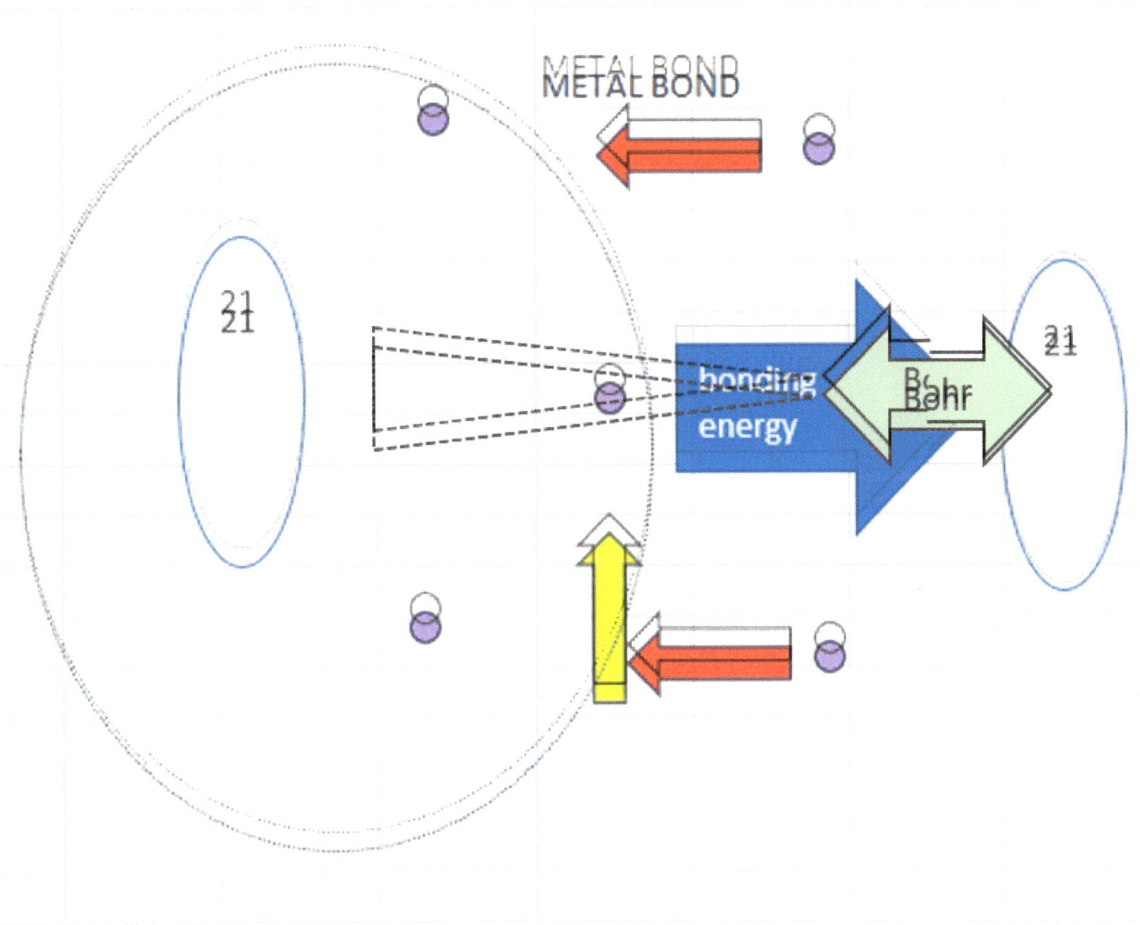

Every negative valance is a 3D channel that is an energy well with very strong walls (yellow arrows) that accept outer electrons of other atoms (if the e-e repulsion is not greater). This is a chemistry and chemical engineering model with which I program and engineer.

So, the system is defined by the electrostatic force of surrounding electrons, plus the excess electrostatic force beyond the equilibrium (modified Bohr radius for O2-He) distance. All particles want

43

to be in the two hemisphere pairs at the 02-He modified Bohr radius, but the outer electrons are beyond that, so they settle into these 'energy wells' in 3D.

All particles have tiny (Heisenberg) ranges of constant pendulums (or better described as Compton effects or O. Consa helical paths). The electrons finds a place based upon other electrons repulsions, but it also has a tiny area in that 3D energy well where it can 'pendulum' or more precisely move in a tiny helical path on the surface of a very small toroid that sits on the larger 3-force toroid.

Even if we have a static model, that would have a settling position. However, there are not iron bars holding particles in one place. Any passing photon or change to nearby molecules gives the electron a little nudge. However, these forces are strong. That creates a pendulum (I refuse to call it an 'orbit' to avoid confusion with the Bohr model; the electron is not going around the nucleus) spiraling within the space as those forces effectively have the electron bounce off the walls (or roll around).

(59) (Scale exaggerated)

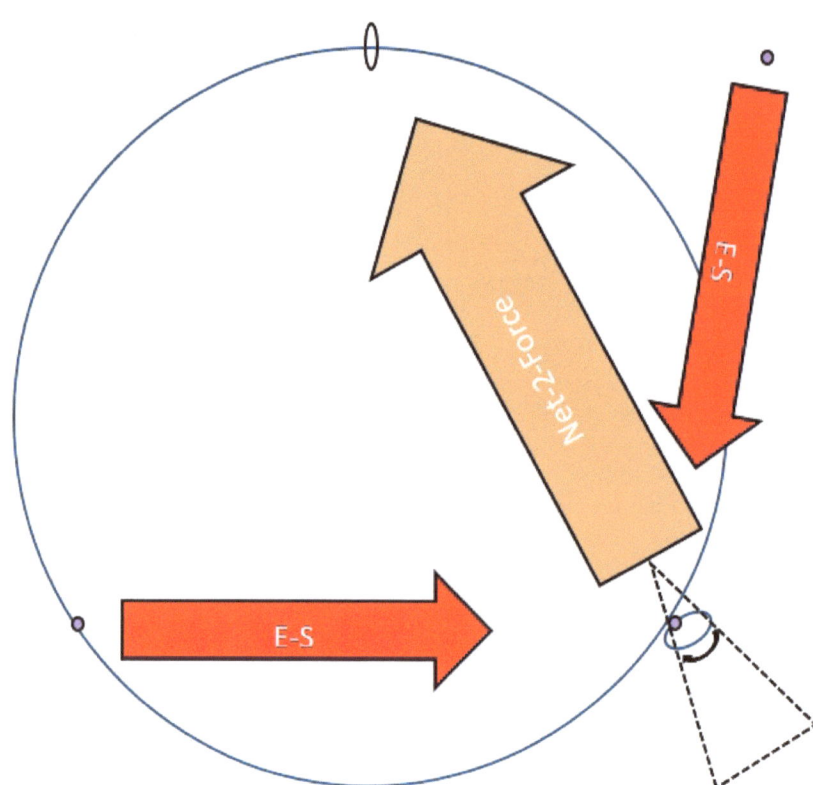

With this, one can see that the Dirac φ is not the correct 3D direction for the freedom of movement. However, the function works in the abstract world of quantum theory. Heisenberg, Compton, and Consa are correct that there is a freedom of movement, and that pendulum is more like a 3D path.

At proper scale, the walls of this energy well, is just that. Like a person in a well 1 meter wide and 10,000 meters deep. (Of course, that walls are virtual, so a passing photon make the calc complex.)

(60)

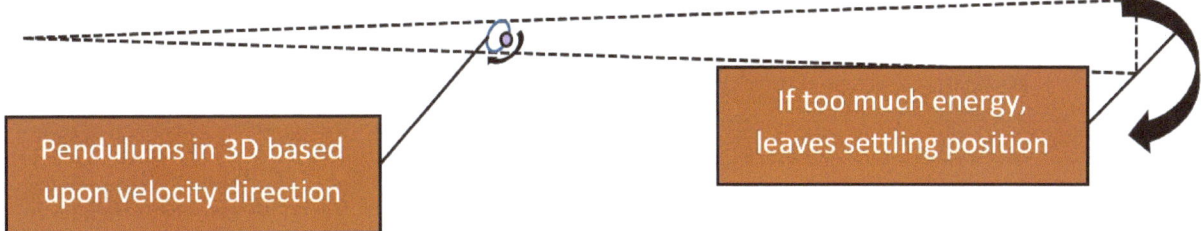

So:

- The walls of the 'energy well' are huge. Electrons generally do not move much. They have reasonably stable positions (crystals, solid, bonding angles).

- There is some movement within that based upon position and momentum (speed + direction) to create tiny 3D pendulum (Heisenberg) which might be a more twisted helical path around that energy well (Compton).

- If you get enough energy, in the correct direction, the electron can escape the 'energy well' and the bond gets broken.

This pendulum (more around a toroid on a toroid, really) of movement is like those fun house putting coins in a conical shape. It does not jump out, but it does 'pendulum' around only until friction (which is not a factor for subatomic interactions) slows that to fall into the middle. (The electron does not 'fall into the nucleus' or even 'fall into the static settling position'; it just keeps operating as a pendulum or spinning top).

However, that is must more complex than the herein static model presented so far. I needs to add the physical dimensions of the particle to get differential forces for that tiny different in distance. It needs to add the right-hand rule. These tiny paths are the major work in quantum mechanics – but without 3D engineering.

However, the above is a sample in that direction for further work that the nucleostaticmagnetics model makes possible, and it will get covered in later chapters. The 3D model of a regular toroid drives engineering that can solve something as complex as the Compton observations of tiny wobbles in electrons or photons. It puts chemistry into clear 3D engineering, and will lead to a Newtonian model to solve all the strange and amazing behaviors that is the subatomic (quantum) world.

So, the overall linear vectors combination on the particle is:

(61)

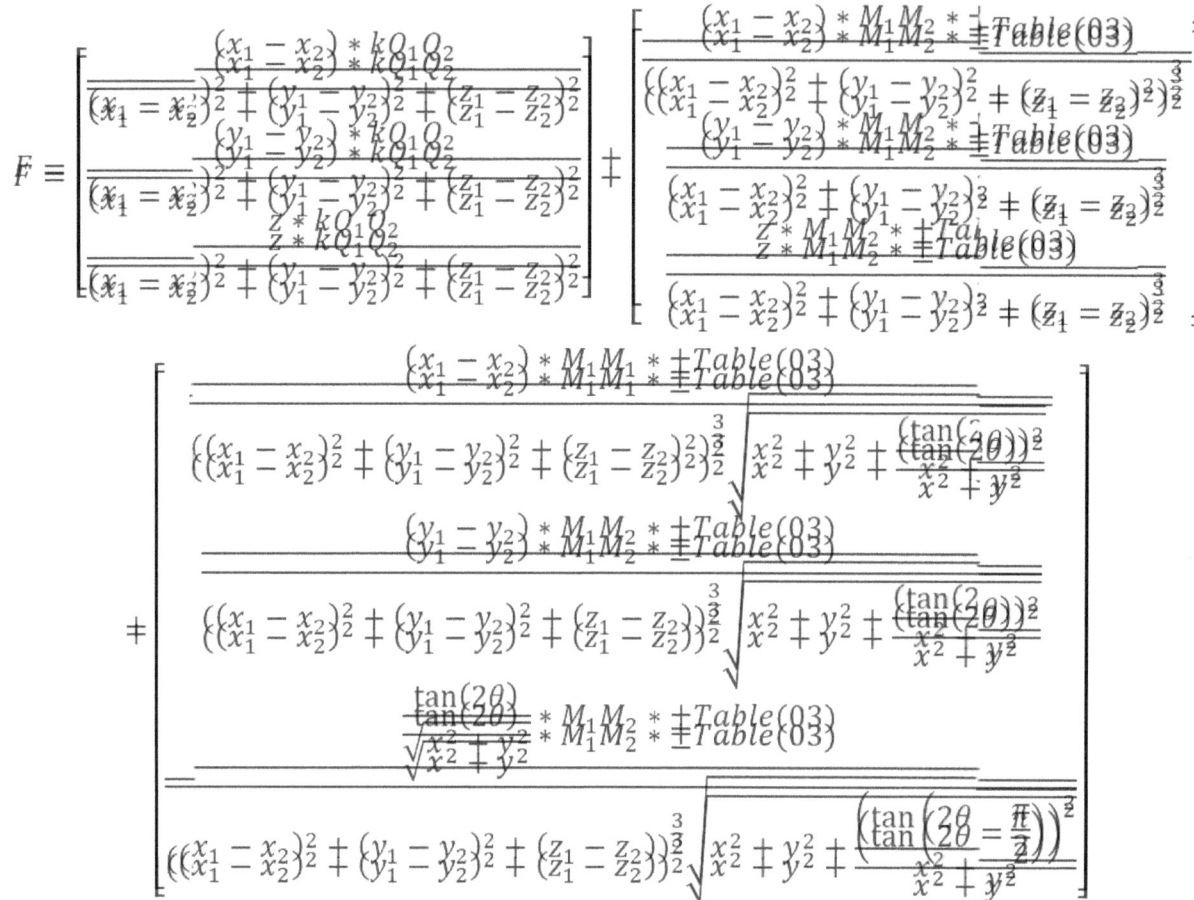

This has probably overwhelmed most readers. It is heavy in the math, and not even complete.

So, before I dive much deeper (sorry, there is still the time-dependent right-hand rule balancing force equation (Maxwell), the multi-particle (chemistry), and the Lorentz field speed-of-light propagation refinements to explain), I will give an overview how this 3-force model already solves the basics of chemistry and subatomic physics.

With the above static model, one can understand a) almost all of chemistry, and b) almost all of quantum theory. That happens because both have the added 3D engineering to provide a frame-of-reference to many calculations that were only statistical (quantum) or only table driven without a correct 3D engineering, equations, and 3D atomic model.

1/Distance-Cubed as Only at Subatomic Distances, Hence, the Wording 'Nuclear'

First, let's review 1/distance-squared ($1/d^2$) vs 1/distance cubed ($1/d^3$) as related to the wording 'nuclear' in the name. These forces are within the atom, near the nucleus, only. Well, 1/distance-cubed gets smaller faster. That is, at increased distances, a force that 1/distance-cubed become immaterial. The 1/distance-cubed gets smaller exponentially faster.

(62)

Relative Distance	$1/d^2$	$1/d^3$	Σ	Relative % of $1/d^2$
1x	1.000000000	1.000000000	2.000000000	50.000%
2x	0.250000000	0.125000000	0.375000000	33.333%
5x	0.040000000	0.008000000	0.048000000	16.667%
10x	0.010000000	0.001000000	0.011000000	9.091%
100x	0.000100000	0.000001000	0.000101000	0.990%
1,000x	0.000001000	0.000000001	0.000001001	0.100%
10,000x	0.000000010	0.000000000	0.000000010	0.010%

So, if one is measuring electrostatic charge at 1 nanometer away, the component of any subatomic $1/d^3$ force would be immaterial (even smaller than above on the relative % on the table). For the more likely setting of a microscope at a centimeter distance, the only observation would be the electrostatic (1/distance-squared) force (with strong and weak off the chart immaterial at 0.000000001%).

Postulate Part 4 – The Strong and Weak Nuclear Forces both operate at 1/distance-cubed ($1/d^3$):

By making both forces as $1/d^3$, we have resolved the limited application range issue for both. They are $1/d^3$, so that means the must be only subatomic (nuclear) distances. The word 'nuclear' has meaning. Further, these two forces do exist everywhere, but the forces are immaterial where people said they "did not exist". In that way, we have eliminated that forces that appear and disappear. They are consistent everywhere, just immaterial at large distances. We can include them, but that would be a waste of time in something like the Closed Container Ideal Gas Law. However, that attribute and clarification is important to **mathematics integrity**. Immaterial is not start-and-stop. So, the postulated strong and weak force equations here are continuous at every distance. We do not have forces that start and stop. We just have two forces that are immaterial at distances beyond the atomic radius, and everyone, including scientists, ignores those two appropriately.

Direct Nucleostaticmagnetics Versus Electrostatic as Electron Field Equilibrium – Why Electrons Stay in a Field!

First, I will focus on strong (direct nucleostaticmagnetics) force. It is the easier to explain.

Now, what happens with direct particle-particles forces when a $1/d^2$ and the $1/d^3$. It is physics magic.

First, the direction or magnitude of direct nucleostaticmagnetics (strong) force is the opposite of electrostatic. That is when one is attractive, the other is repulsive. Newton and the universe prefer the equilibriums.

For this step, we will focus on the one dimensions of distance. That both forces are directly related to distance, and thereby isotropic (the same in every direction). That fun iron-filing shape will come later in weak force, so bear with me.

So, I present the physics graph of force strength versus distance only for a) electrostatic at 1/distance-squared ($1/d^2$), and b) direct nucleostaticmagnetics (strong nuclear) force at 1/distance-cubed ($1/d^3$) for the nucleus to electrons interaction. Now, electrostatic is protons and direct nucleostaticmagnetics (strong nuclear force) is nucleons, is protons + neutrons, but that would just change the scale, so I have focused on 01-H Hydrogen as that has not neutrons.

The below uses the concept of **physics-positive** forces as **repulsive**; the interacting particle force **increases (+) their relative distance**. It also uses **physics-negative** as **attractive**; the interacting particle force **decreases (-) their relative distance**. Now, that is acceleration, not velocity, but most people understand it better without the 'decrease' stated as 'decelerates'. That way the signs follow the math of the acceleration logic given the position and velocity as the other inputs.

In this presentation, the electrostatic force in attractive (opposites attract) is always physics-negative attractive. It is stronger the closer the particles get, and weaker as distance (the 'X' axis increase). This force continues going to negative infinity (the red line). However, direct nucleostaticmagnetics (strong nuclear force) is repulsive (physics-positive everywhere). It (blue line) decreases steeper at close distance (inside the balancing =1 at the modified Bohr radius), and its distance from the axis is less at great distance than equilibrium.

So, the net-2-force is green, and that a) has an equilibrium (which postulate as the Bohr radius – more later); b) is repulsive inside that radius, and c) is attractive outside that radius.

(63)

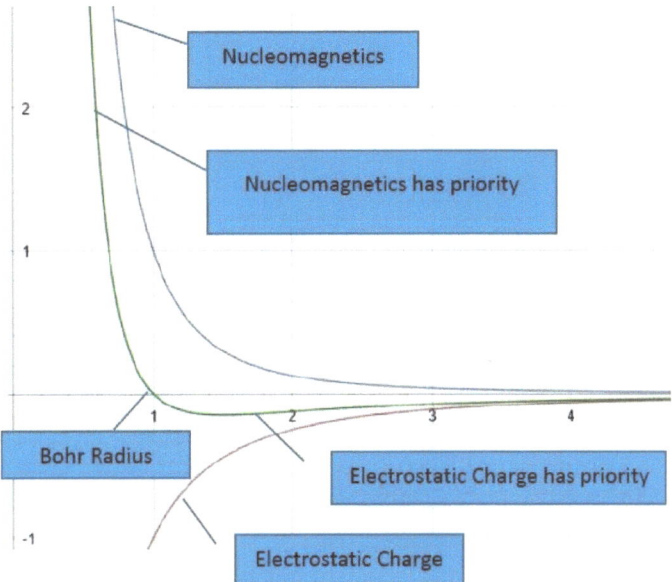

A math-physics downslope through zero is magic. A self-balancing equilibrium. Physics magic.

Electrons to not fall into the nucleus. The electrons build at an average distance that is consistent. The combination of 1/distance-squared isotropic electrostatic and the Newtonian offsetting 1/distance-cubed axial, anisotropic nucleostaticmagnetics force becomes the fundament balance that create electron is fields/shells.

That electrostatic force alone (red line) will go to negative infinity. However, that does not happen. It is the first of the strange features of the subatomic world. Instead, the 1/distance-cubed nucleostaticmagnetics fully offsets at the equilibrium.

However, that offset has the first strange result. As one gets toward zero, a distance of 1/x becomes **infinity**. That is:

(64)

$$x > \infty \text{ as } 1/x > 0, \text{ so}$$

$$\text{As } x => 1/\infty \text{ or } 0, \text{ Electrostatic} > (\infty)^2 \text{ AND Nucleostaticmagnetics} > (\infty)^3$$

Infinity-cubed $(\infty)^3$ is bigger than infinity-squared $(\infty)^2$. That one in mind-wrenching, but has math integrity. Most interestingly, nucleostaticmagnetics becomes infinity-cubed repulsion from nucleon to electron, but the electrostatic only becomes infinity-squared as an attraction. Cubed beats squared, even at infinity, so the electron is net repulsed inside the nucleus!

So, as distance gets smaller, inside that equilibrium (modified Bohr radius), the **nucleostaticmagnetics forces becomes the most important force**. The use of electron-Volt in electrostatic is wrong when inside the atom. It is not electrostatic force at all that directs and drives subatomic particle behavior.

Also, that means that particle inside the atom (distance) act based upon magnetics more. Ah, the behavior of quantum theory, that molecules act strangely inside the nucleus, but follow electrostatic perfectly outside the nucleus. That concept will drive all the later postulates.

The net-2-force is the core physics and math hereafter. It is an entirely new way of thinking about forces, and thereby a new physics. However, it is that simple graph. I can explain it to non-PhDs.

The red line is the same as the green line everyone beyond atomic distances, but inside, one must use net-2-force. For centuries, one would never measure nucleostaticmagnetics directly. Only the behaviors indicate its existence and properties.

When a net-force crosses zero in a distance graph going down, that is a physics magic balancing point:

- At the zero net-2-force distance, a particle will not move (if it has no momentum!)
- A particle drifting to greater distance than that will get attracted back (green line is physics-negative)
- A particle getting too close (smaller distance), the electron will get repelled outward (green line is physics-negative) to the equilibrium

This is self-correcting system. That is the magic of atomic physics when the proper equation gets applied.

Further, in a time study, that system will have pendulums! Ah! So, quantum theory is correct; there are these energy pendulums of electrons in harmonics of position and momentum. Quantum mechanics is correct, but now has a physic model underlying it. All of Heisenberg is about tiny pendulums of position trading with momentum. That is, electrons move in 3D engineering wells at one of many pendulums.

(65)

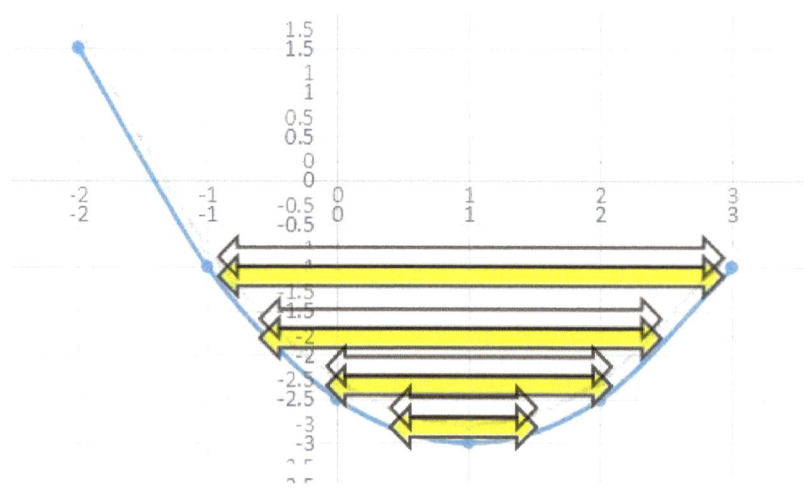

These are 3D engineering concepts of net-2-force (electrostatic + strong) electron-nucleus attraction as 'down' in the above, and electron-electron repulsions as the 'walls' at the side creating the energy wells.

Particles can be in any pendulum cycle (unknowable) from the variety (knowable) in yellow. Each gives difference changes of position and velocity (momentum) that is combination are **the pendulums** defined (by electrostatic force and often its cousin gravity, both isotropic 1/distance-squared force). That unknown pendulum with a knowable set is the probability math that creates quantum mechanics. All of quantum mechanics is the conversion of many possible pendulums (in that interaction of position and velocity/momentum running back and forth within a defined energy well) into a e-exponent (e^x) prediction.

Now, in many calculations you can ignore one or the other of the two forces – they are nuclear only. At 1,000 layers of molecules, any nucleostaticmagnetics force is immaterial to the outside world (1/1000)-cubed; that is 0.000000001x.

That is why scientists have missed the above for a century. That variance is immaterial to a microscope even a millimeter away. A microscope cannot measure the variance caused by the extra nucleostaticmagnetics force at that distance.

Postulate Part 5 – The Strong Nuclear Force is repulsive between nucleons (both protons and neutrons) and electrons. That creates the equilibrium which is the shells/fields. That is the opposite direction as electron-proton electrostatic attraction. *The application of strong force to nucleus-electrons is novel.*

So, what is the static equilibrium? When do the force calculations balance? For electrostatic:

(66)

Factor	Charge-Force Estimation
Charge-Force Constant (Coulomb)	$k_e = 10^{10}$ [8.99×10^9 m³ kg²/ (s²)]
Charge 001-H Hydrogen atom = 1 proton	$q_1 = 10^{-19}$ [1.67×10^{-19}]
Charge 001-H Hydrogen atom = 1 proton	$q_2 = 10^{-19}$ [1.67×10^{-19}]
Distance	d=5.29×10^{-11} m or 10^{-10} m
	(d²=5.29×10^{-11})² m = 10^{-21} m
Exponent shortcut	+k+Q+Q-d-d
Short-cut calculation	10-19-19-(-21)= 10-19-19+21= 10-38+21=-7
Charge-Force Attraction	10^{-7} m¹/ (s²) (Newtons)

Yet, the nucleostaticmagnetics (strong force) calc is the same strength at the Bohr radius:

(67)

Factors	Magnetic-Force Estimation
Charge-Force Constant (Coulomb)	$M_A = 10^{-38}$ m³ kg²/ (s²)]
Charge 001-H Hydrogen atom = one (1) Proton	n=1 or 10^0
Charge 001-H Hydrogen atom = one (1) Proton	n=1 or 10^0
Distance	d=5.29×10^{-11} m or 10^{-10} m
	(d³=5.29×10^{-11})³ m = 10^{-31} m
Exponent shortcut	+k+Q+Q-d-d-d
Short-cut calculation	-38+0+0-(-31)= -38+0+0+31= -38+0+31=-7 N (Newtons)
Charge-Force Repulsion	10^{-7} m¹/ (s²) (Newtons)

Here is the math:

For 01-H Hydrogen, the force balance. There is one proton in the nucleus, and one electron, so easy.

(68)

Proton Charge	Nucleons	Electron Charge	COMBO	COMBO	Distance	squared 8.2464436016144E+25	cubed 4.3638300230498E+15	(partL)*m*s^(-2)*(part1,2)^(-2)
01-H Hydrogen (zero Kelvin)								
1	1	-1	-1	1	5.2917721067000E-11	-2.9448584712019E+46	2.9448584712019E+46	0.0000000000000E+00 (partL)*m*s^(-2)

For 02-He Helium, notice that I use the same constants for every Element and molecule. Also, the extra neutrons change the nucleus count applicable to the Direct N-M calculation. Further, notice that nucleostaticmagnetics does not apply (red blank space) for the electron-electron interaction.

(69)

Proton Charge	Nucleons	Electron Charge	COMBO	COMBO	Distance	E-S squared	Direct N-M cubed	Net	
02-He Helium (zero Kelvin)									
2	4	-1	-2	4	1.2095479101029E-10	-1.1273286335070E+46	9.8641255431859E+45	-1.4091607918837E+45	(partL)*m*s^(-2)
2	4	-1	-2	4					(partL)*m*s^(-2)
-1	1	-1	1	1	2.4190958202057E-10	1.4091607918837E+45		1.4091607918837E+45	(partL)*m*s^(-2)
					Net for on electron 1m1 (1s1)			0.0000000000000E+00	0.00000000%

It works beyond Shell-1, but the electron-nucleus-electron pairs need a tight 3D structure, so the math is more complex, and gets presented in my separate papers.

So, the postulated physics equation has an equilibrium, that that is the same equation that we will use for direct nucleostaticmagnetics (strong nuclear) force at every particle combination and distance. Math integrity.

The Breakthrough Varying from Statistical Methods - Planck

Over a century ago, Max Planck found the oscillation of quanta. His postulate changed a graph from infinitely increasing temperature, to match the experimental results for the black body experiment. A red line to infinite was replaced with a supported graph of a hump attenuating to zero. The original was good at the left, Planck matched everywhere.

(70)

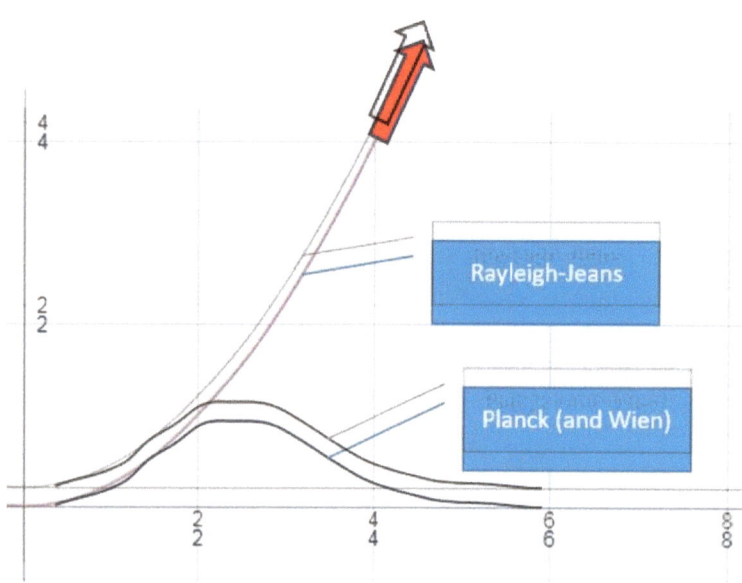

While Rayleigh-Jeans (red line) was excellent to a point. Yet, eventually, it fails; it keeps going to positive infinity wrongly. Planck explained that with the same type of reasoning as applied here. Classical mechanics only had the one force, electrostatic, a red arrow going to infinity.

Yet, today, the current variations of solution of the Schrodinger Equation all use only electrostatic force. The Distribution Function Theory from Hartree and Kohn-Sham through current work of Perdew and K. Burke (UC-Irvine) all focus only on electrostatic, 1/distance-squared calculation. They instead postulate 'holes' in these probabilities. However, none of those avoid discontinuities. Hartree equation works because it is focused on gas state with the rotational energy creating the probability normalizing, but its further application to bonding in solids has the same design error as Rayleigh-Jeans.

Today, electrostatic-only calculations work great outside the electron-settling distance, but fail miserably inside the atom. It is the infinity problem in the opposite direction, negative infinity of the nuclear proton to the free electron. It needed an entirely new approach. In the below, the electron-proton electrostatic goes to negative infinity (red line), but scientific observations are to me the green line. In the same way, the net-2-force is a breakthrough that replaces electrostatic-only which is a red line to physics-negative infinity (attractive) into a model where there is equilibrium to generate a static model for electrons settling into a field.

54

There is an electrostatic-only (red arrow) going to infinity (negative attractive in this case), but here is an offset, an 'oscillation'[xi] of the 'magnetic field'[xii] (the blue line of nucleostaticmagnetics) that now comes to its direct relationship for particles to particles. Those words are from a time-dependent model. Here we define the static relationships first (before adding time-dependent models). That generates the net-2-force (electrostatic + strong/direct nucleostaticmagnetics) for the nucleus electron relationship as the green line:

(71)

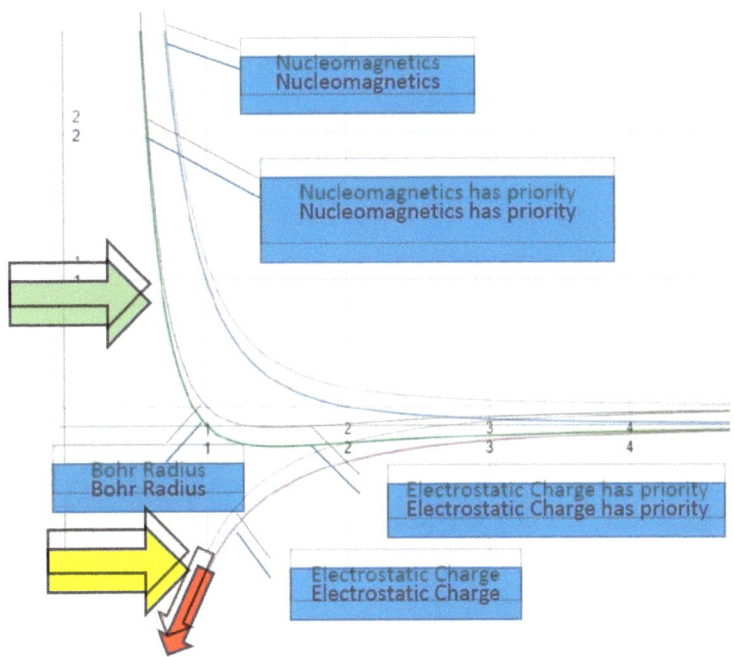

The red electrostatic-only equation gets replaces with an equilibrium where electron do not fall into the nucleus (yellow arrow). The net-2-force (green) uses an offsetting strong nuclear force called direct nucleostaticmagnetics (blue line) here.

Restating Strong Force as Direct Nucleostaticmagnetics of Proton-Neutron-Proton 3D Structure

That last section was the strong force applied to nucleus-electron (as repulsive balance creating the shells/fields). However, most textbooks and research focus on strong nuclear force as the binding that keeps the nucleus together as a unit. That is where the wording 'strong' occurs. Protons would repel – with nuclear explosion force if only protons with electrostatic force (like-kind repel). My postulate that strong force also applies nucleus-electron is novel. So, part of the **mathematical integrity** and continuity is that the same formula creates nucleus binding force (strong nuclear) force. The same force of the last section is the same calculated for inside the nucleus.

Postulate Part 6 – The direct nucleostaticmagnetics (strong nuclear) Force is attractive between nucleons (protons or neutrons) and other nucleons (protons or neutrons). That creates the binding in the nucleus structure, <u>so long as a neutron settles into a 3D physical position between every proton</u>.

This same force is attractive between nucleus which the offset the proton-proton electrostatic repulsion (like-kind repel). In fact, that repulsion is huge given the distances are smaller. This is really inside the atomic shells, so this nucleostaticmagnetics become overwhelmingly important. Here is the start of understanding magnetics. The secret of the nucleus staying together is the neutron, a particle with nucleostaticmagnetics, but no charge, that magnetic force stay strong in a chain:

(72)

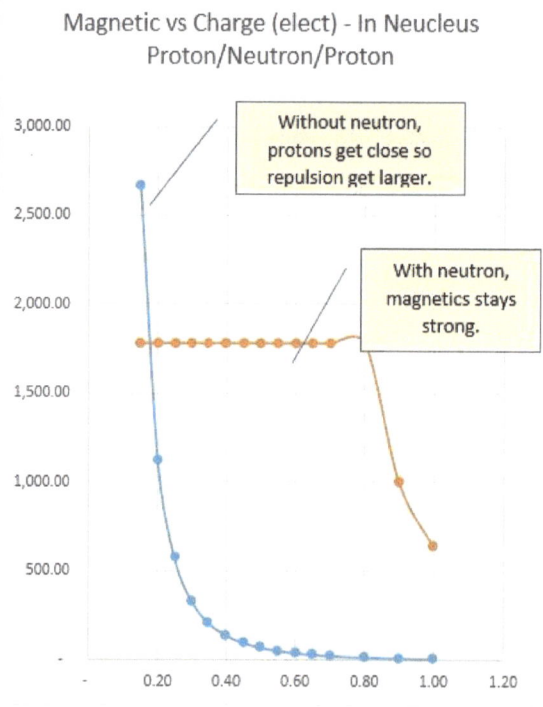

However, here I need to explain that the current textbooks do not present the above graph, but instead present a 3-particle system (without their intention). Here is the graph of the strong force.

Take a look at the existing observed shape of the strong nuclear force:

(73)

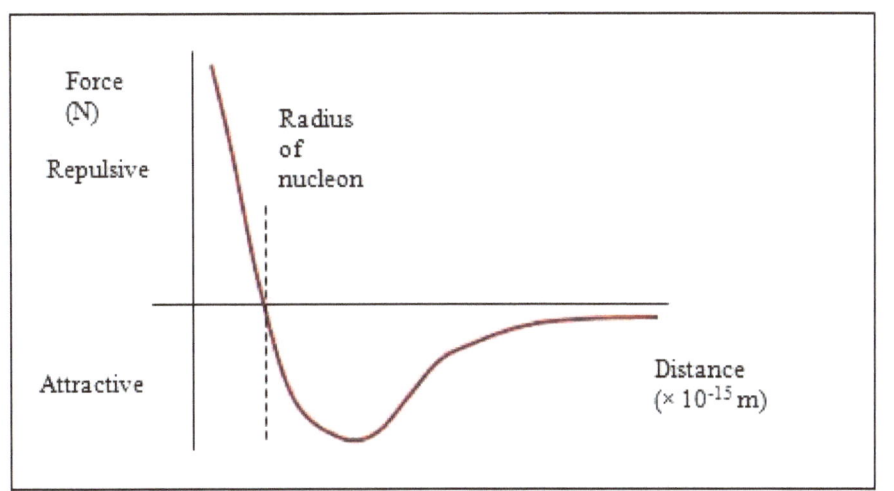

xiii

Now review the shape of the force curve when you have nucleostaticmagnetics-only at distance 1x for a neutron, and both nucleostaticmagnetics and electrons-proton-proton repulsion as distance of 2x.

(74)

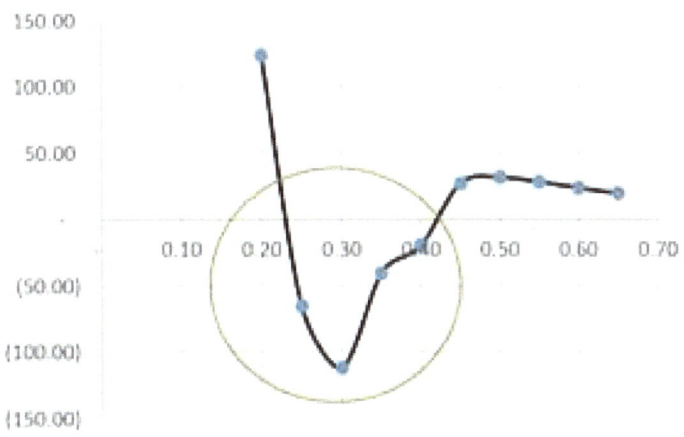

The math of that simplified example is:

(75)

								Cubic	Square
	Obviously, these are all 10E-17 distance:					Cubic		Blue	Orange
	I took of that factor to make the graph show.					Blue	9.0000	1.6020
		Cubic	Square				9.0000		1.6020
						Magnetic		Elect-Charge	0.05
		Displacement (separation by neutron)						Stays strong	
	x	Factor1	Factor2	x	y1		x	y2	
	0.01	1,000,000.00	10,000.00	0.01	9,000,000.00		0.01	1,780.00	
	0.02	125,000.00	2,500.00	0.02	1,125,000.00		0.02	1,780.00	
	0.03	37,037.04	1,111.11	0.03	333,333.33		0.03	1,780.00	
	0.04	15,625.00	625.00	0.04	140,625.00		0.04	1,780.00	
	0.05	8,000.00	400.00	0.05	72,000.00		0.05	1,780.00	
	0.06	4,629.63	277.78	0.06	41,666.67		0.06	1,780.00	
	0.08	1,953.13	156.25	0.08	17,578.13		0.08	1,780.00	
	0.10	1,000.00	100.00	0.10	9,000.00		0.10	1,780.00	
	0.15	296.30	44.44	0.15	2,666.67		0.15	1,780.00	
	0.20	125.00	25.00	0.20	1,125.00		0.20	1,780.00	
	0.25	64.00	16.00	0.25	576.00		0.25	1,780.00	
	0.30	37.04	11.11	0.30	333.33		0.30	1,780.00	
	0.35	23.32	8.16	0.35	209.91		0.35	1,780.00	
	0.40	15.63	6.25	0.40	140.63		0.40	1,780.00	
	0.45	10.97	4.94	0.45	98.77		0.45	1,780.00	
	0.50	8.00	4.00	0.50	72.00		0.50	1,780.00	
	0.55	6.01	3.31	0.55	54.09		0.55	1,780.00	
	0.60	4.63	2.78	0.60	41.67		0.60	1,780.00	
	0.65	3.64	2.37	0.65	32.77		0.65	1,780.00	

The shape is the same in the critical range (yellow circle). A N-M neutron will hole two E-S+N-M protons together in a knowable physics calculation with a N-M neutron between. The neutron attracts both neutrons, and that is more than the repulsion of the E-S proton-proton because that distance is at least 3x more = (a radius + a diameter) versus just less than a radius, or even touching.

Remember that the direct nucleostaticmagnetics (strong) force is continuous, touching at 1/distance-cubed for distances much less than 1 (the He Bohr radius equivalent). As a result, the magnetics holds stronger if locked in place, than the proton-proton electrostatic repulsion at the P-N-P separation.

Now, the current presentation is multi-particle, the P-N-P set, not just the proton<>neutron. That is what creates that double hitch in its shape. I do not present the base equation as with the bottom-hump, then upper-hump attenuating back to zero. What I present is that is the three-particle (or more) interaction. So, the postulate is that the strong force presented is confusing because of the double-hump section. However, that is understandable if one presents the same direct-nucleostaticmagnetics for multiple-particle sets. One gets a double-hump by calculating with the neutron at $d=r_{neu}$ and with protons at $d=(2r_z^{+r_{neu}})$.

The part at the right of my drawing, the part going to physics-positive does not apply because that would be the next neutron-proton set. That repulsive section (physics-positive) would not occur if the structure was P-N-P-N-P.

That leads to a 3D engineering model of a nucleus:

- Proton-Proton Electrostatic while huge, gets reduced by the spacing of a neutron in between.
- The nucleostaticmagnetics force is larger because a) the neutron is touching protons and/or neutrons around it so the strength is greater because of less distance separation, and b) magnetic field stay strong, and even aggregate in a chain.

That is, for a nucleus to remain stable, in 3D engineering, there must be a neutron locked into a position.

Well, the evidence is strong. In simple molecules, the ratio is 1P:1N up to ~20 particles. Probably a chain or a ring. However, then you need extra neutrons for a stable. The structure gets complex in 3D to a) keep a neutron between each layer.

First, a simple chain or ring where each proton is separate by a neutron. They have a magnetic axis that links the set, so it remains physically stable. The protons (red spheres) each have a neutron (green sphere) creating separation. Further, the entire structure works with the poles of each particle connecting = locking in the structure.

(76)

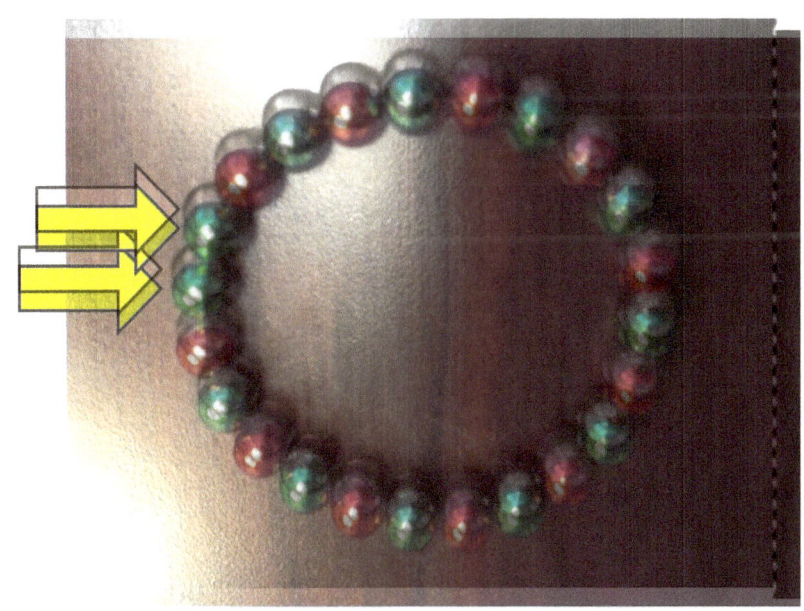

Notice that at the left, a position exists with two neutrons. That is an **isotope** (extra neutron) because you get different configuration with extra neutrons, but that still keeps all protons separated in 3D.

However, it more complex when you get more than 20 protons. Instead, you need more than one neutron in 3D to keep the structure stable as protons can touch in any direction. One needs a neutron in multiple directions for different potential proton-proton interaction that would break the nucleus structure.

(77)

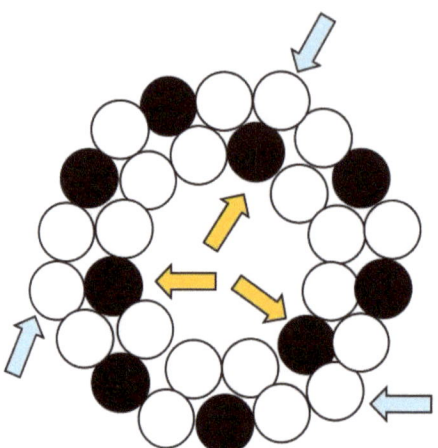

For each proton (black spheres at yellow arrows) one needs an extra neutron (blue arrows) as the structure builds in layers. In cannot be 1P:1N.

Instead, you get exactly what happens in evidence. The number of neutrons a) starts at 1:1, then grows at a faster rate. Finally, the rate at the end is 3:1 which is like a tetrahedron to get a new proton on a pedestal locked by 3 neutrons below it.

(78)

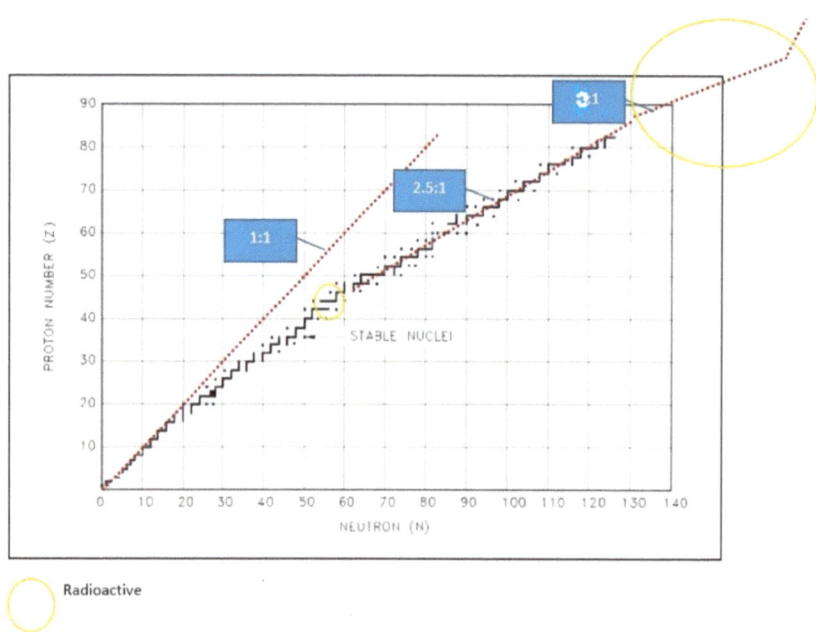

That pedestal for radioactive, very-large Elements is a triangular pyramid (a tetrahedron) at a 3N:1P ratio:

(79)

One can see the ratio of 3N:1P from Lead to Uranium in the graphic (20). The 3:1 ratio follows from 3D engineering.

That leaves the calculation of proton-proton repulsion with the distance of a neutron.

If the neutron is missing, the distance is less than the proton radius, as they almost touch, and the force 10^{+8}:

(80)

Factor	Charge-Force Estimation
Charge-Force Constant (Coulomb)	$k_e = 10^{10}$ [8.99×10^9 m³ kg²/ (s²)]
Charge Proton	$q_1 = 10^{-15}$ [0.9×10^{-15}]
Charge Proton	$q_2 = 10^{-15}$ [0.9×10^{-15}]
Distance	$d = 3.0 \times 10^{-16}$ m or 10^{-16} m
Exponent shortcut	+k+Q+Q-d-d
Short-cut calculation	10-16-16-(-10)-(-10)= 10-16-16+15+15= 10-32+30=8 N (Newtons)
Charge-Force Repulsion	10^{+8} m¹/ (s²) (Newtons)

The protons, as above that are near touching, will decay and some subset of particles will fly off at nuclear explosion acceleration.

However, if that distance increase to If the neutron gets added, the distance 1-1/2 particle radius units away, and the force 10^{+2}:

(81)

Factor	Charge-Force Estimation
Charge-Force Constant (Coulomb)	$k_e = 10^{10}$ [8.99×10^9 m^3 kg^2/(s^2)]
Charge 001-H Hydrogen atom = 1 proton	$q_1 = 10^{-19}$ [1.67×10^{-19}]
Charge 001-H Hydrogen atom = 1 proton	$q_2 = 10^{-19}$ [1.67×10^{-19}]
Distance	d=0.86 × 10^{-15} m or 10^{-15} m
Exponent shortcut	+k+Q+Q-d-d
Short-cut calculation	10-19-19-(-15)-(-15)= 10-19-19+15+15= 10-38+30=+2 N (Newtons)
Charge-Force Repulsion	10^{+2} m$_1$/(s^2) (Newtons)

However, the direct nucleostaticmagnetics (strong nuclear) force would calculate as stronger because its distances is still that touching distance (must less plus **so strong** based upon 1/distance-cubed!)

(82)

Factors	Magnetic-Force Estimation
Charge-Force Constant (Coulomb)	$M_A = 10^{-38}$ m^3 kg^2/(s^2)]
Nucleon 001-H Hydrogen atom	Z=1 or 10^0
Nucleon 001-H Hydrogen atom	Z=1 or 10^0
Distance	d=0.9 × 10^{-15} m or 10^{-15} m
Exponent shortcut	+k+Q+Q-d-d-d
Short-cut calculation	-38+0+0-(-15)-(-15)-(-15)= -38+0+0+15+15+15= -38+45=+7 N (Newtons)
Magnetic-Force Attraction	10^{+7} m$_1$/(s^2) (Newtons)

There it is: The same force equation working for direct nucleostaticmagnetics for a) electrons at the shells/fields of the atom and b) the nucleus at the center of the atom. Math integrity:

Strong Force at Direct Nucleostaticmagnetics for Electron Distribution – The Start of Magnetics

Now, we start to introduce magnetics. The first part is field strength, it changes between the poles and the axis. I was slightly incomplete before, only focusing on the denominator, the 1/distance-cubed ($1/d^3$). Now, we will address strength based upon inclination angle, the start of the magnet part.

Further, this step requires the combination of 1/distance-cubed force: direct nucleostaticmagnetics (strong nuclear) and axial nucleostaticmagnetics (weak nuclear) forces.

The strength of direct (strong) and axial (weak) nucleostaticmagnetics varies based upon the inclination angle.[xiv] It is stronger at the equator, and weaker at the poles (I will explain how different form a macro-world magnet later).

(83)

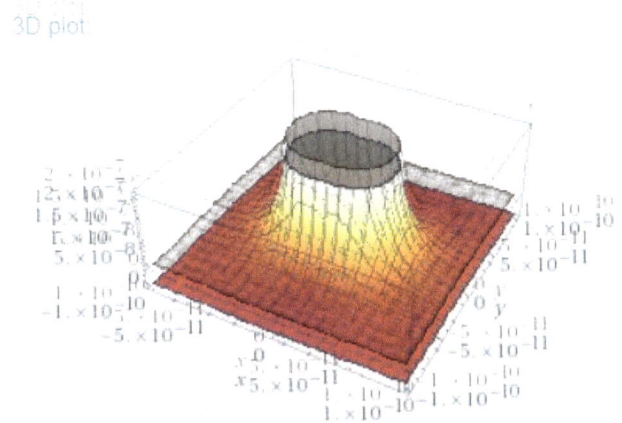

3D plot:

The above is a 2D slice of a 01-H Hydrogen atom, with the 'z' direction as the axis direction. It is enormously repulsive if you get to close, with valleys at the two poles. There at the poles, the net-2-force strength is than zero attractive!

The first two electrons want to settle at the poles. That matches the Periodic Table of Elements. The first shell has 2 electrons, then full.

Here is the above as an electron distribution. Again, different than the 'fuzzy ball' of the current solutions of Schrödinger for 01-H Hydrogen or 02-He Helium.

(84)

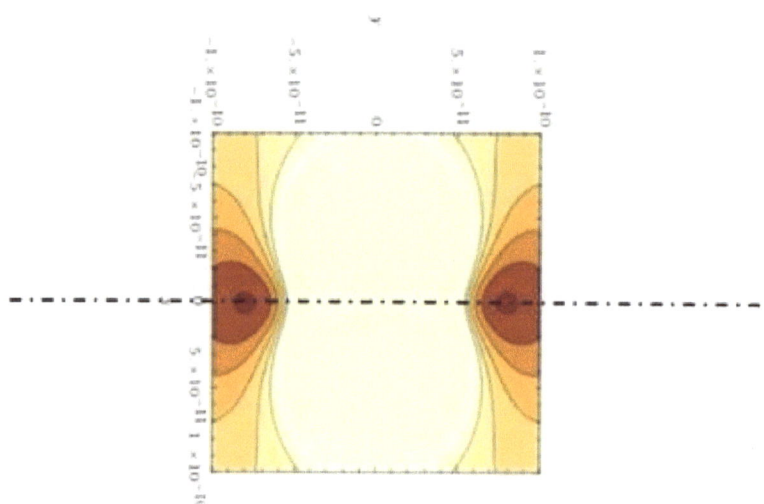

Postulate Part 7 – Valid solutions to the Schrodinger Equation need revision to including an inclination angle factor which creates two nodes in two (2) hemispheres, and Pauli pairs need to get evaluated in 3D as a locked set of electron-nucleus-electron with the electrons at 180-degrees with the nucleus as the vertex.

These are energy wells, valleys, that drive electrons to either of the two poles. In 01-H Hydrogen, the one polar position fills to balance the one proton in the nucleus, and the second is the position for hydrogen bonds. In 02-He Helium, both positions fill to match the two protons in the nucleus, and everything is full. So, Helium is unreactive, and a noble gas – no places to bond.

The current model is a fuzzy ball, without electrons falling into the nucleus. However, this model is much better. Electrons do not fall into the nucleus . . .

But, the observation is from a random angle, like the entire structure rotating as in gas state. That means the distance is sometimes the full, and sometimes at an angle, smaller. On average, they might appear as that fuzzy ball. The resulting distances to protons and electrons from the above equation will give the same results as the fuzzy ball of the current solution.

So,

The direct nucleostaticmagnetics atomic model then presents that 'strong nuclear' force as a) nuclear binding, **and b) electron-shell repulsive** and c) immaterial beyond that. However, the force is everywhere. Further, the equation never changes, the force physics are consistent, determinant, Newtonian, and vary only with the particles and distances involved. Math integrity.

Weak Force at Axial Nucleostaticmagnetics for Electron Distribution – The Basis of Pauli ½ and Quantum Numbers

Remember that this model has two hemispheres.

The two hemispheres become Pauli pairs and Pauli's +½ and −½. Those are hemispheres, so they have half the total energy.

With the application of direct nucleostaticmagnetics (strong nuclear) force as repulsive to an equilibrium, that means that Pauli pairs settle at the 180-degree positions. The math is that electrons want to get as far away from each other as possible, but still get to that equilibrium position relative to the nucleus. So, you end with electron-nucleus-electron as locked sets.

(85)

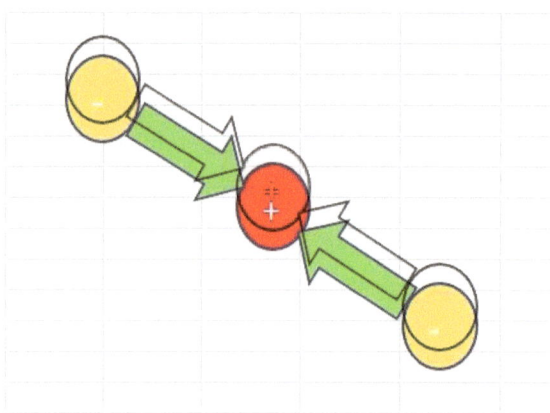

That means they operate where one goes 'up', the other 'down'. The nucleus becomes the vertex or fulcrum creating this dynamic. (Sorry, the graphic is right vs left not up versus down.)

(86)

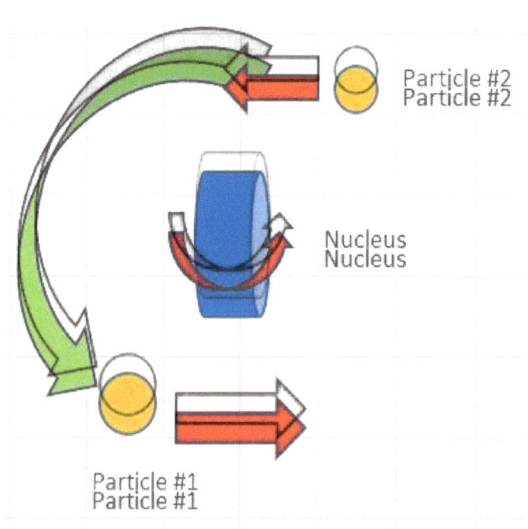

Postulate Part 8 – The Pauli ½ and pair concepts should get restated for electron-nucleus-electrons linked sets in 3D engineering.

This approach is further clear by re-evaluating the equation hc/eμ = 2 at (9) in Dirac's 1931 paper.[xv] If one takes the postulated hemisphere approach, then in any delivery of energy, the Planck-Einstein equation portion gets corrected that the delivery is only ½ of the energy, so 'hc' becomes Planck-Einstein-Pauli 'hcm$_s$' with m$_s$ as the Pauli +½ or -½. That segment becomes is 'hc/2'.

This has profound impact. The =2 in that Dirac (9) equation creates the discontinuity between the macro-world and the subatomic (quantum) world. There is always a jump from the denominator to the numerator. However, as I change the equation to =1, then the subatomic world has a direct equality to the macro-world. As a result, the macro-world and the subatomic-quantum world do not have a discontinuity.

Postulate Part 9 – The Planck-Einstein Equation should get restated for subatomic interaction to understand that the delivery of energy can be only one hemisphere. So, it becomes Planck-Einstein-Pauli:

(87)
$$E = hcm_s = \pm \frac{hc}{2}$$

This is, **a) two pulses for each whole particle**. Each rotation is the passing of two energy wells, two pulses . . . not one (1).

Further, this is **b)** the 3D engineering concept that **all wave functions operate from 3D particle rotation** . . . in those hemisphere units. It is rotation, but a rotation that hits the minimum (less repulsion) twice as often. That subatomic particles are not **monopoles**, but **duopoles**.

(88)

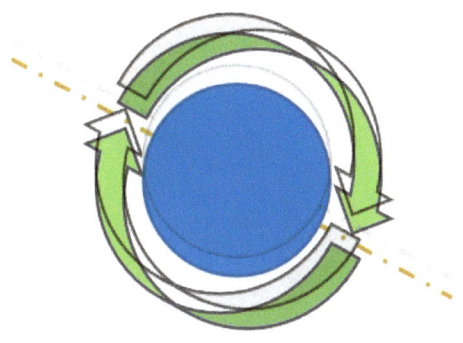

Postulate Part 10 – The Dirac 1931 paper at (9) should get restated for subatomic interaction to understand that the delivery of energy in subatomic-quantum systems can be only one hemisphere of energy in each cycle. So, it becomes:

(89)

$$\frac{\frac{hc}{2}}{e\mu_0} = 1$$

By the way, that fixed Dirac 1931 (9) to equal 1, so that the quantum world and macro-world link 1:1. Please note the critical nature that lack of discontinuity at (9), then breaks the requirement at his (3) for the quantum method applied. The math conversion to e^f requires that the position and momentum are unknowable. That issue is not applicable as this paper does not need the quantum mechanics methods.

Further, this provides 3D engineering solutions such that quantum mechanics and quantum theory have physical meaning, using a 3D model of subatomic particles as a particle settling positions relative to the nucleostaticmagnetics axis of the nucleus. That means that you have a) layers, b) inclination/longitude, c) latitude/azimuth, and finally which hemisphere. That is, the Quantum Number relate to 3D engineering in 3D space and 1D time. Here is the static model:

- 1st AVSC **Layer**/Shell Quantum Number

- 2nd AVSC **Inclination**/Longitude Quantum Number for the count of the subshell from the polar axis filling in structures of squares 1/3/5/7/7/5/3/1 from pole to equator to pole which become 1+3+5 = 9, the cube of 3, in each hemisphere, so 2x9=18 total electrons.

- 3rd AVSC **Latitude**/Azimuth Quantum Number for the count of electron at the same latitude in each hemisphere based upon sizing of 2nd Inclination Quantum Number

- 4th AVSC **Hemisphere** Quantum Number of +1/2 and -1/2 of that hemisphere

(90)

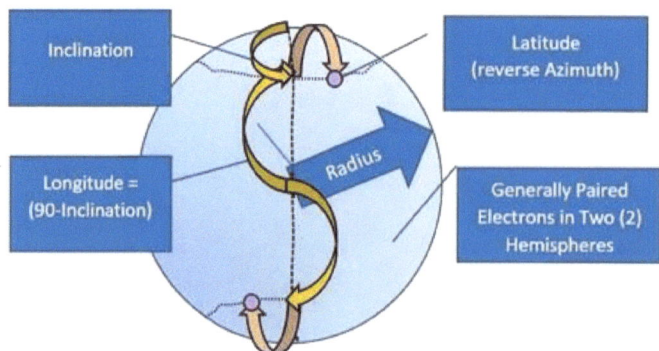

AVSC Number Model – Spherical with Hemispheres

The direct relation of Dirac's (9) as adjusted for Planck-Einstein-Pauli, instead of just Planck-Einstein, becomes a function that has position, velocity, and acceleration/force. The divergence of the quantum-world and macro-world at Dirac's (9) goes away as that ratio goes to 1. A '1' is a direct correlation.

Beyond the abstract, the direct relationship has physical 3D meaning. Further, that means that quantum mechanics (QM) remains fully valid. That is every equation that works in quantum theory will work here. However, we can get more as we will show in the end.

Axial Nucleostaticmagnetics for Electron Distribution – The Basic of the Periodic Table in Anisotropic Strength of Strong and Towards the Axis Magnet-Like Feature of Weak Force

Now, the magnetics really comes into play. We can extend this logic and 3D engineering into building of the Periodic Table of Elements.

Postulate Part 11 – Full Shells in the Periodic Table of Elements are based upon two hemispheres in the 'z' dimension, and the tightest configuration in 3D, a circle, in the remaining two dimensions. So, $2 \times \pi r$-squared. That is interlaced as the hemispheres can fit tightest with a 2nd layer offset by ½ phase. That is, the even shells locked by 0-degrees latitude/azimuth, and the even shells by 180-degrees latitude/azimuth.

This strength weak at the poles creates a structure where the poles fill first. That is just like the Periodic Table where the first two Columns are Alkali Earth and have specific properties. Further, that generates the 3D engineering for every subshell as a set of electrons at the same energy level by being at the same inclination, starting with the two (2) poles by 1 electron in each shell.

This hemisphere (Pauli's ½) approach provides the logic of the subshells and the full shells.

The engineering is that shells are 1/3/5/7/7/5/3/1 from poles to equator to pole. That means that full shells are two (2) hemispheres by PI*r-squared in unit jumps for the two remaining dimensions.

(91)

- Shell 1: $2 \times 1^2 = 2 \times 1 = 2$ (as one subshell located at the zeroth position from the polar axis)
- Shell 2: $2 \times 2^2 = 2 \times 4$ (which is 1+3) = 8 electrons
- Shell 3: $2 \times 2^2 = 2 \times 4$ (which is 1+3) = 8 electrons offset by a half phase to tightest packing
- Shell 4: $2 \times 3^2 = 2 \times 9$ (which is 1+3+5 as 3 subshells) = 18 electrons
- Shell 5: $2 \times 3^2 = 2 \times 9 = 18$ electrons offset by a half phase to tightest packing from Shell-4
- Shell 6: $2 \times 4^2 = 2 \times 16$ (which is 1+3+5+7 as 4 subshells from pole to equator) = 32
- Shell 7: $2 \times 4^2 = 2 \times 16 = 32$ electrons offset by a half phase to tightest packing from Shell-6

That leaves the equator view as the building of larger electron longitude/azimuth sets starting with one at the poles, increasing count as they get to the largest seven near the equator.

(92)

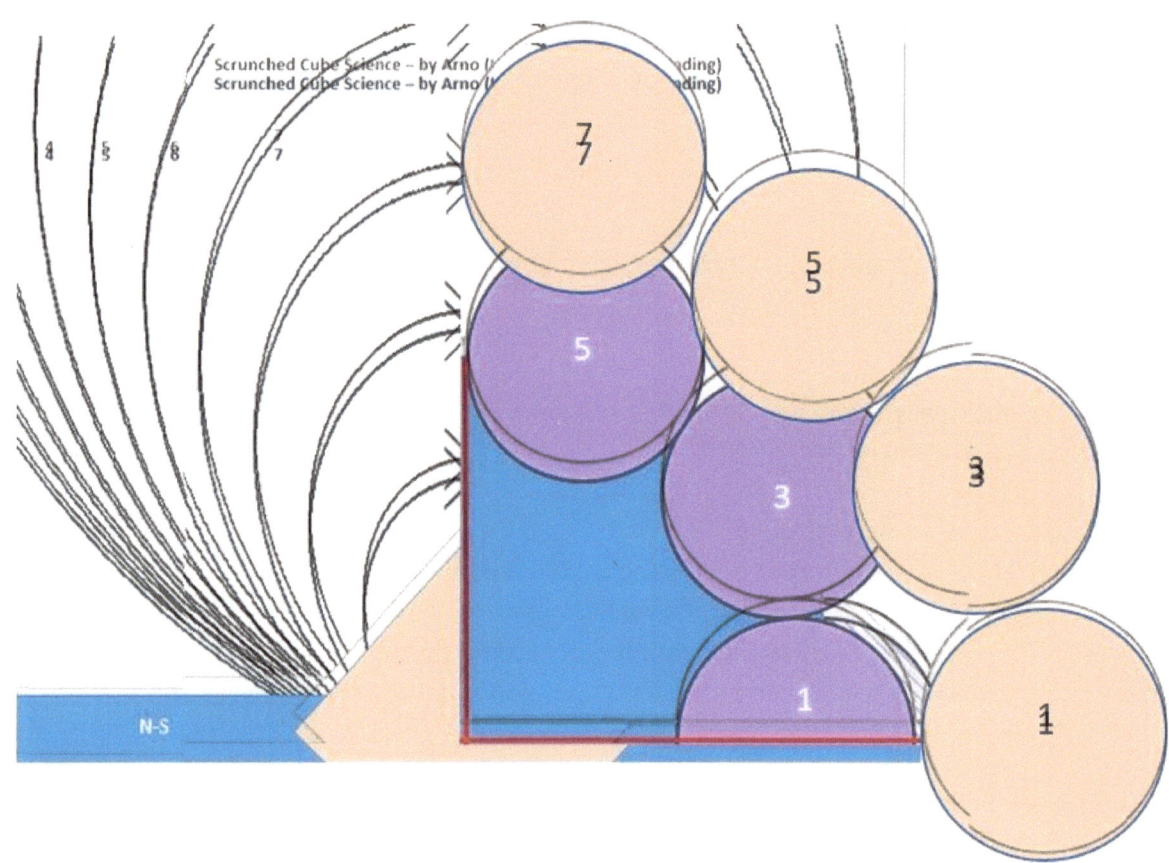

Obviously, the first shells can fit only 1 x 2, one at each pole of two hemispheres. The second and third shell fits 4 x 2, the 4th and 5th fits 9 x 2 electrons. The π is in every question, so that disappears in comparison.

In the largest sense, this is a strict Aufbau filling order*. So, that scientific observation remains herein.

* Again, there is a variance from the current presentation that comes separately, in a=my paper just on exceptions to aufbau.

So, quantum number become workable engineering. However, as described later, this is not strict, and only applies in the full subshells and shells, and gets adjusted by bonding and alloy process. As a results, the work is much richer, increasing the need to this 3D net-2-force modelling. One cannot just follow Aufbau, nor the gross hemisphere and squares interlaces above. As documented late the results are much more complex when integrating both 3D force orientations and time.

This is spherical coordinates: radius, inclination/longitude, and latitude/azimuth, with the extra factor that these build in two hemispheres.

(93)

- First Layer Number ($r_\#$) - the number of the layers from the nucleus
- Second Inclination/Longitude Number (as I present $\theta_\#$) - the number of the groups from the pole
- First Latitude/Azimuth Number (as I present $\phi_\#$) - the number of the electrons in the set at that inclination
- First Hemisphere Number (m_s or I present as $\zeta_\#$) - as $+\frac{1}{2}$ or $-\frac{1}{2}$

Notice that I change the names to add the '#' because the concepts are with 3D, and not just quantum abstract concepts. They are counts of particles in the 3D reference framework of the four parts.

Postulate Part 12 – The Periodic Table of Elements gets build from the poles (weakest repulsion) to the equator. The use of integers is the physics count of particles, and the ½ also has a physical meaning.

I have two books about Chemistry, and the Periodic Table, as the topic itself is so broad. As such, I shall not repeat that work here.

Axial Nucleostaticmagnetics Rules in Application – Linear and Rotational

First, one must understand that this is free particle physics. Particle can move (linear) and particles can rotate. By the way, rotations express a wave function (a key element of quantum mechanics).

Towards-the-axis of free particles creates different types of force:

- For the particle at a distance from the axis, this is a linear force towards-the-axis

- For a particle at the fulcrum of the axis, that is, on the axis, the force is rotation to move its axis to point toward the other particle.

(94)

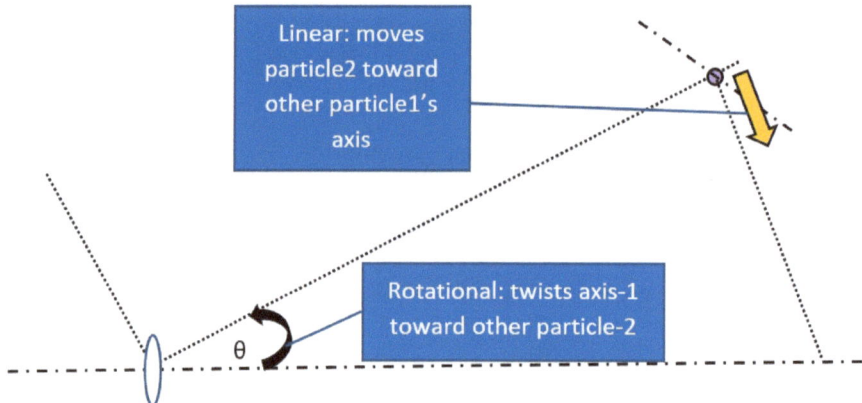

Unfortunately, that is only the impact of the nucleus, in this case, nucleostaticmagnetics field of the nucleus. However, there is also the nucleostaticmagnetics field of the electron, that have the reverse settings:

- The electron's axial nucleostaticmagnetics interaction want to move (linear force) the nucleus towards its axis.

- The electron itself has a force to rotate its axis towards the direction of the nucleus.

So, each particle has a linear (particle) force and a rotation (wave) force. Wave-particle duality is a natural result that two subatomic particles interact with different ways to express.

(95)

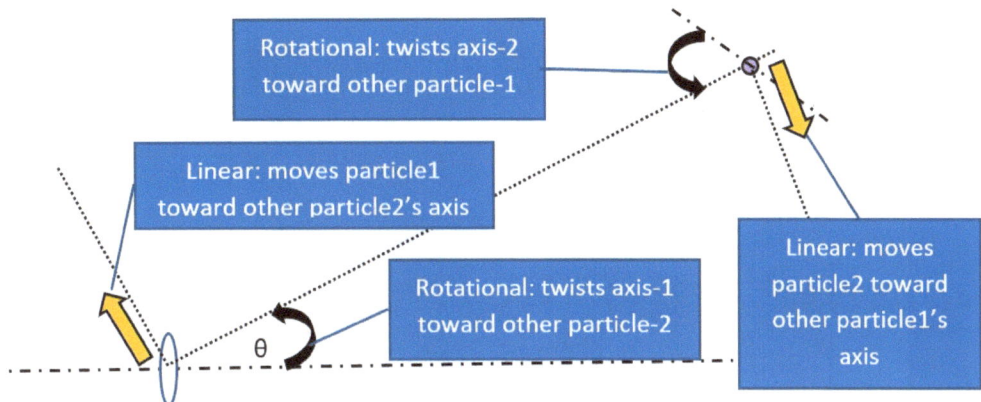

So, I focus on the 4th interaction; the electrons want to rotate towards the nucleus. That would be great at near zero Kelvin, the electron would stop rotating. However, now we get to position and velocity. We are creating a 'spinning top':

- The electron was to rotate its axis toward the nucleus

- The electron has freedom of moment preference in the φ spherical orientation.

Every time the electron rotates, and it passes beyond the axis, and keeps going. It gets into a pattern (with other electrons and the nucleus), that becomes the angular moment specific to that system.

(96)

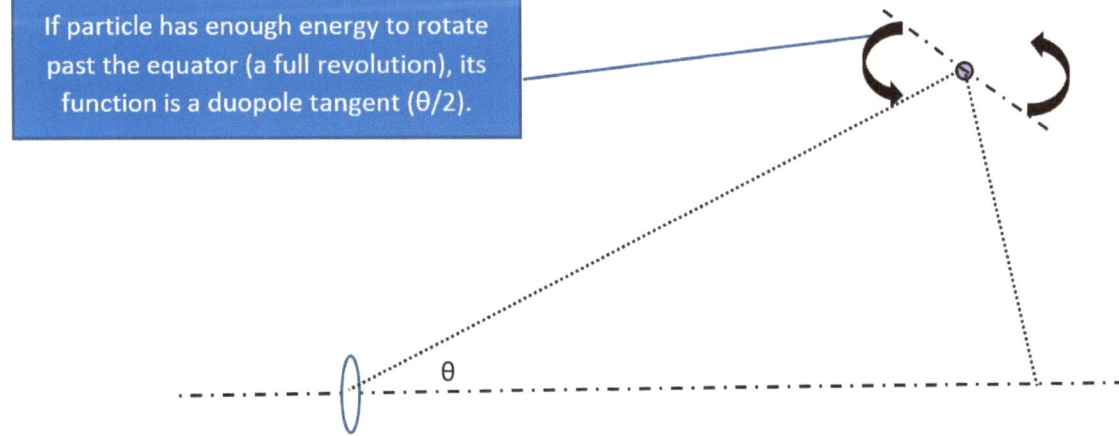

The energy as the axis passes to the other hemisphere takes that great leap from negative infinity to positive infinity. Yet, the particle does not leap out of the electron, as those forces get overwhelmed by the time and rotational momentum.

Postulate Part 13 – The average of E-S field tangent and the N-M towards the axis combine for a axial nucleostaticmagnetics (weak nuclear) force generates the Dirac 4th Equation when enough energy to rotate the electron's (not the nucleus's) nucleostaticmagnetics axis beyond the equator relative to the nucleus.

The above only evaluated the electron, but the full system has additional layers of complexity:

- How does the electron Pauli-hemisphere pair impact the above vectors?
- The nucleus (nucleons as a set) have the reciprocal linear and rotational force, so what happens there?
- There are other electro-nucleus-electron sets at different angles, so what happens for sets at different inclinations and different counts. What is the spherical drum?

The easiest one: This 3D engineering has an odd count of electron in one hemisphere and the same number in the opposite hemisphere offset by ½ phase.

That makes the polar view of the linear forces on the nucleus as offsetting. These, from the polar view are at 0, 120, and 240 degrees, and completely offset (as they are the same strength). This means that in the x,y or longitude orientation, there is not new force on a full electron shell.*

* All of chemistry is about the combination of non-full electron shells, and in my Chemistry textbook. That is more expansive than it can get covered in this paper.

(97)

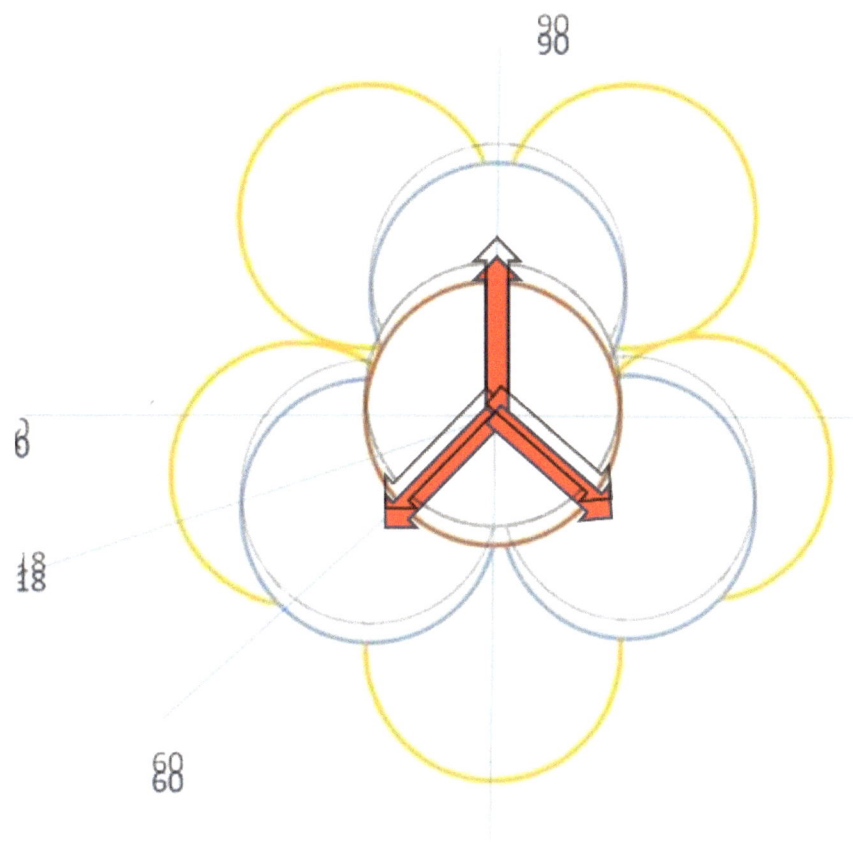

The opposite hemisphere is the same, just upside down at 60, 180, and 300 degrees.

Similarly, any portion of the vector in the z- direction from one inclination would get fully offset by the z-direction component of the other hemisphere vectors when combined.

Postulate Part 14 – That nucleus does not have linear motion because the set of same latitude forces offset.

As similar 3D engineering also occurs for the nucleus structure regarding rotational forces. As a result, the nucleus wants to rotate towards a) particles evenly along a latitude/azimuth, and b) sets in both hemispheres. That leads to a similar picture, but the forces are rotational, not linear:

(98)

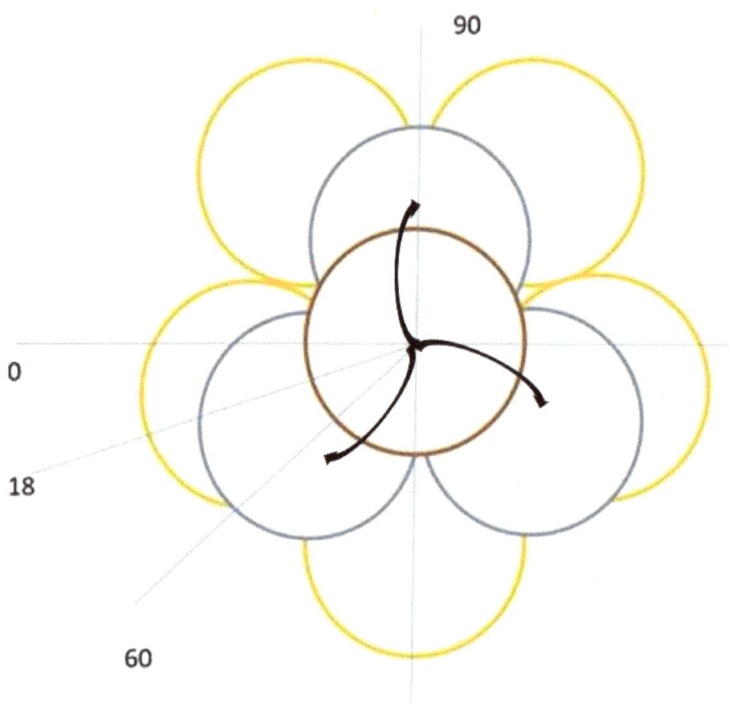

I prefer the equatorial view to understand each fully. Remember that the linear towards the electron's axis, and the rotational force on the nucleus's axis goes towards the electron. There is also offset forces.

(99)

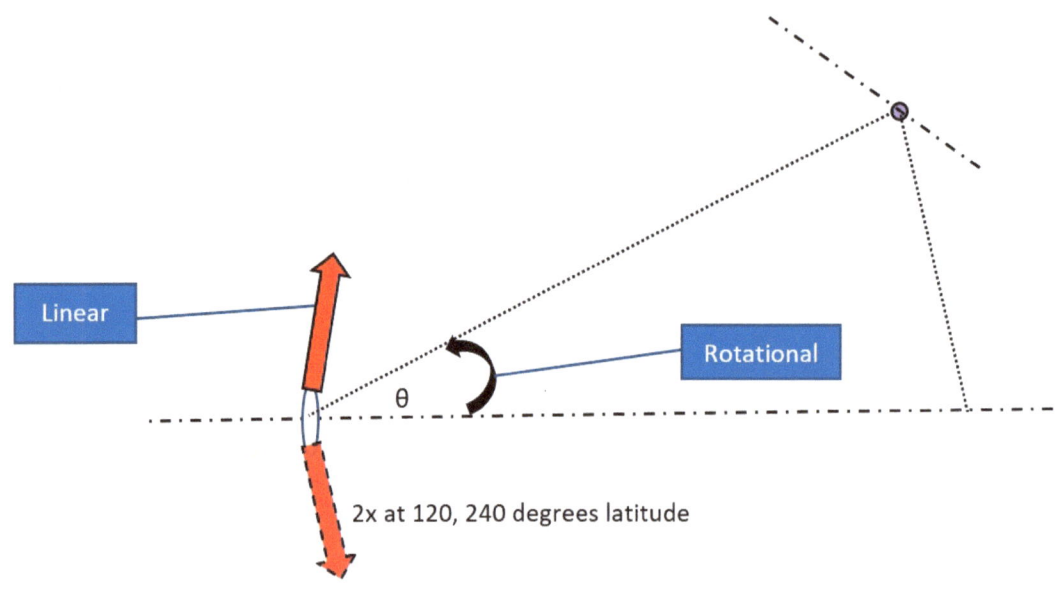

However, there are the full set of particles, so the rotation is also downwards for two other particles.

(100)

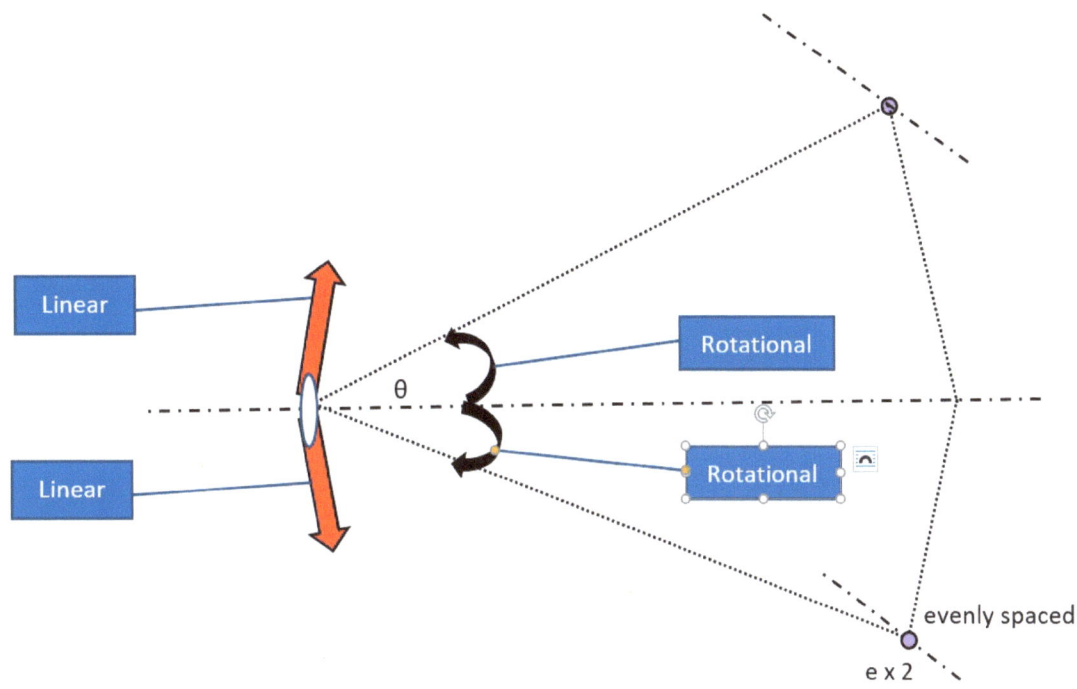

Postulate Part 15 – That nucleus does not have rotational force because the electron of-the-same-hemisphere-subshell positions pull the axis in offsetting directions.

This becomes the most complex issue. All of quantum mechanics and wave functions is that the electron axis is rotating, past the equator. That makes all subatomic physics as a 3D engineering physical model that will generate logic that we need.

Over time, this will come with further papers, but that full scope is beyond this paper.

Multiple Electron-Nucleus-Electron Sets Become a Spherical Drum

The greater challenge is the calculation of a whole molecule or atom with these electron-nucleus-electrons sets at different a) distances, and b) inclination angles. However, all of those must operate such that the nucleus does not get torn apart.

This creates the quantum mechanics concept of a spherical drum. The rotational rate of all these set must get into a harmonic that works for all the forces at different levels (based upon a) and b)).

(101)

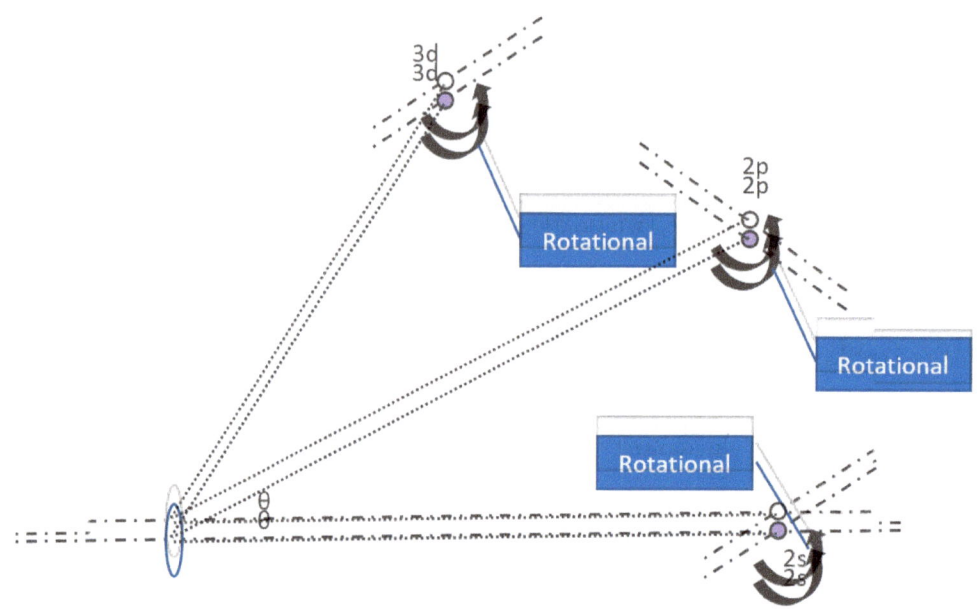

In fact, these rotational wave-functions at different angles and distances (from other subshells), creates a set of rotations. However, those rotations do settle into a synch, and that is the all of Quantum Numbers.

The simple math is that various rotation seeks to get in synch.

(102)

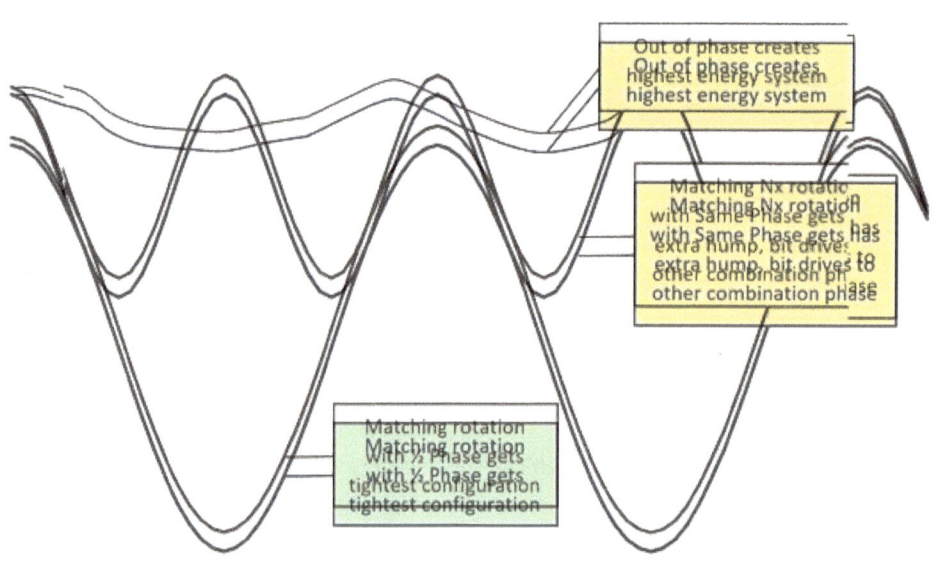

This unit rotation rates specific to a) the electron-nucleus-electron set, and b) the harmonies with the larger set, get driven into this spherical drum, that highly prefers 1x, 2x spherical harmonic drums:

Multiple Fields and Changes Always Creating Counter-Forces (Traditional Magnetism and Lorentz) and Surface Field Differentials

The great challenge is the understanding of fields as reactive in time working with the nature of fields. First, there are two field shapes, the isotropic field of electrostatic force (spherical – red line), and the nucleostaticmagnetics field (the blue toroid/bagel).

(103)

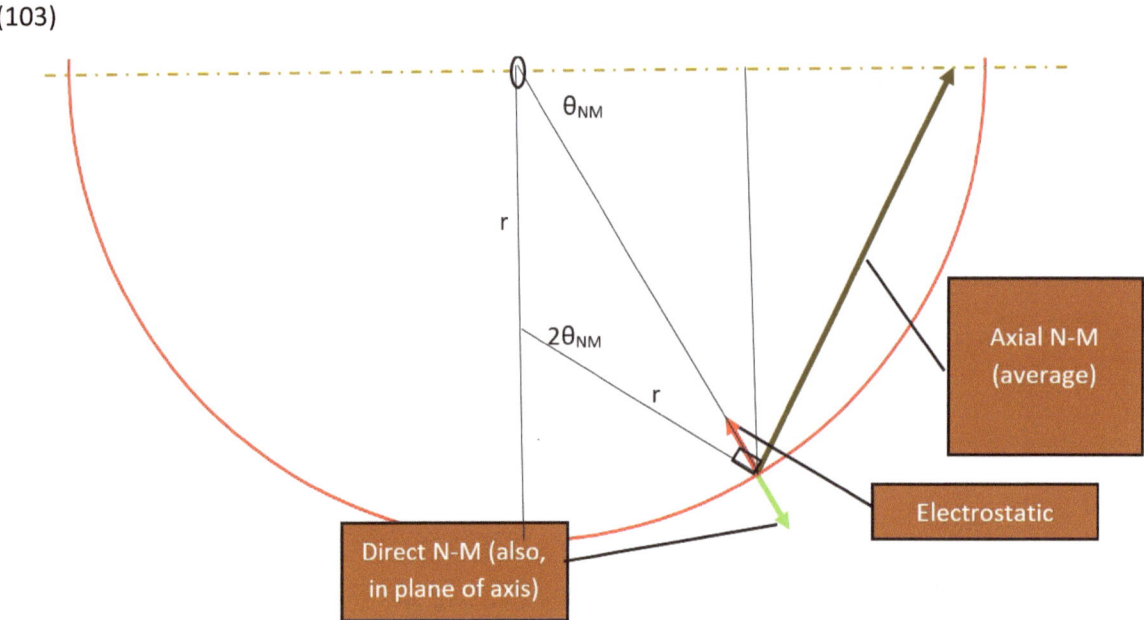

That means whichever direction the particle moves, one of those fields (or both) get a variance, and that creates counter forces (traditional magnetism and/or Lorentz forces).

No matter the action of a particle, one of the fields (electrostatic or nucleostaticmagnetics) is disturbed.

In the above, if the particle moves along the toroid which has the same nucleostaticmagnetics strength, then the electrostatic field changes. If one moves along the nucleostaticmagnetics field, then the electrostatic force changes.

Further, the particles have dimensions, to there is always a surface differential over the side closer or further from the orientation. The close side towards the nucleus

Multiple Inclination Angles Settles into Only Unit Rotation Rates for that Common Set

However, I think the next generation can improve by making the inclination as discrete inclination/longitude angles.

In fact, this is most exciting because one can take a) the speed of light, b) the effective radius of the electron particles, and find that these sets **only work at unit increments**. This is quite complex because the spherical harmonic relay via the nucleus (black double arrows). That is the main part why 1x, 2x. I will explain the rest in a later paper as that itself is a huge topic.

(104)

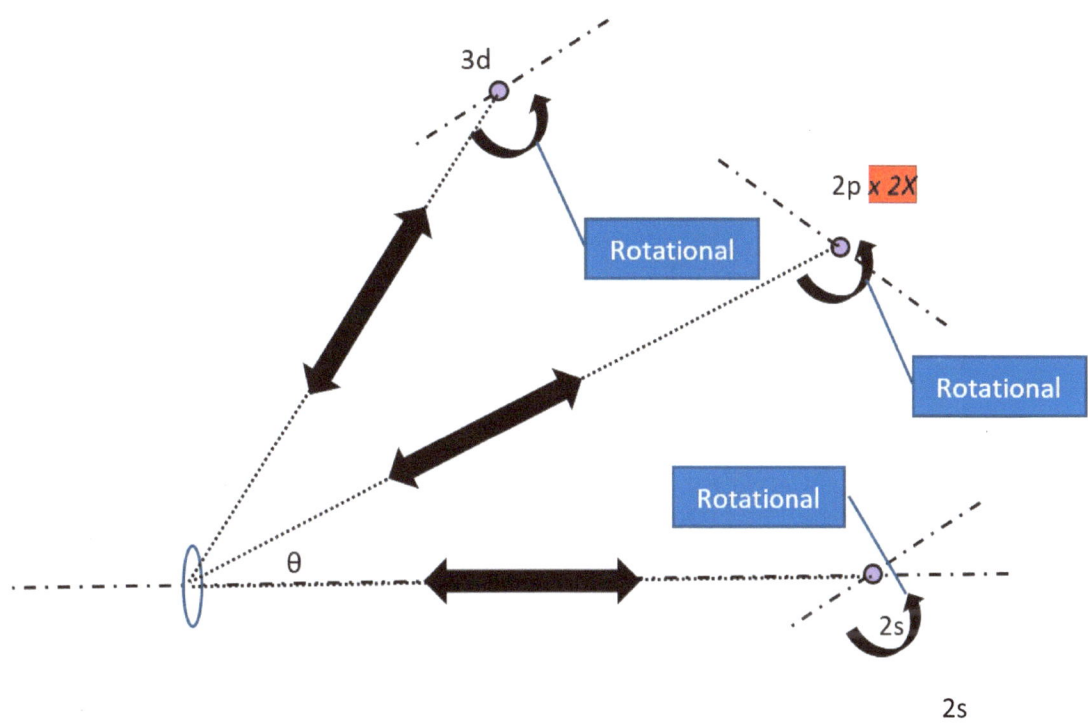

This is the math of all electrons want to get to b) that modified Bohr radius equilibrium distance and b) the nucleostaticmagnetics axis, and c) as far from their 180-Degree Pauli-hemisphere paired electron as possible.

Postulate Part 16 – That presentation about ½θ is about one electron's activity, which relays off the nucleus. However, that creates the entire system becomes a 3D spherical drum with lots of forces on the nucleus.

Rules in Application for Three Fundamental Forces

The three fundamental force operates as an interlinked set, especially within the atomic range. That is because two of the forces decline exponentially more rapidly.

- Electrostatic force is isotropic at 1/distance-squared ($1/d^2$) in the direction between the two particles

- Direct nucleostaticmagnetics (strong nuclear) force strength decrease at 1/distance-squared ($1/d^3$) in the direction between the two particles.

- Axial nucleostaticmagnetics (weak nuclear) force has strength decrease at 1/distance-squared ($1/d^3$) in the direction between the two particles.

Electrostatic rules are based upon Coulomb's Law which is the multiplication of the charges of the two interacting particles. The rules for interactions between particle for electrostatic force is based upon multiplication, and the sign of each particle:

(105)

	Proton (+) Proton (+)	Neutron (0) Neutron (0)	Electron (−)
Proton (+)	+ * + = + Physics-repulsive	+ * 0 = 0 No interaction	+ * − = − Physics-attractive
Neutron (0)	0 * + = 0 No interaction	0 * 0 = 0 No interaction	− * 0 = 0 No interaction
Electron (−)	− * + = − Physics-attractive	− * 0 = 0 No interaction	− * − = + Physics-repulsive

The nucleostaticmagnetics rules for interactions are quite different. Both protons and neutrons act the same. Further, it really is two classes:

- Nucleons (protons or neutrons) interact with each other as attractive (physics-negative)
- Electrons interact with nucleons as repulsive (physics-positive)

Now, since the electrostatic has 3 x 3, I can state with the same matrix.

(106)

	Proton Proton	Neutron Neutron	Electron
Proton	(−) Attractive	(−) Attractive	(+) Repulsive
Neutron	(−) Attractive	(−) Attractive	(+) Repulsive
Electron	(+) Repulsive	(+) Repulsive	No interaction

Future Work: Applications of Toroid Nucleostaticmagnetics Forces and the Right-Hand Rule of Such Activates. Time-Dependent Application for Axial Nucleostaticmagnetics (Weak Nuclear) Force and Magnetic Fields

All the above focused on the static model, and I only have so much space here. However, while there are a time-dependent model. That is, when a particle moves by the direct nucleostaticmagnetics (strong nuclear) force (yellow) or the axial nucleostaticmagnetics (weak nuclear) force (yellow), there is a reactive effect on the overall magnetic field.

In this way, the field balancing is in the 3rd direction – along the regular toroid – for magnetic fields. That one moves on the direct-N-M (strong-green) or the axial-N-M (weak-brown), the magnetic field operates in the right-hand rule out from the paper! The 'X' marks the position of the vector out from the paper.

That Maxwell remains the rule at every distance. Math integrity.

Postulate Part 17 – Maxwell's Equation and time-dependent modeling become continuous at all distances as this :

Please know that the electron in this magnetic field will want to move in the plane of the surface of the regular toroid = perpendicular to the direction of the regular toroid center, and the point in that plane directed towards the axis.

(107)

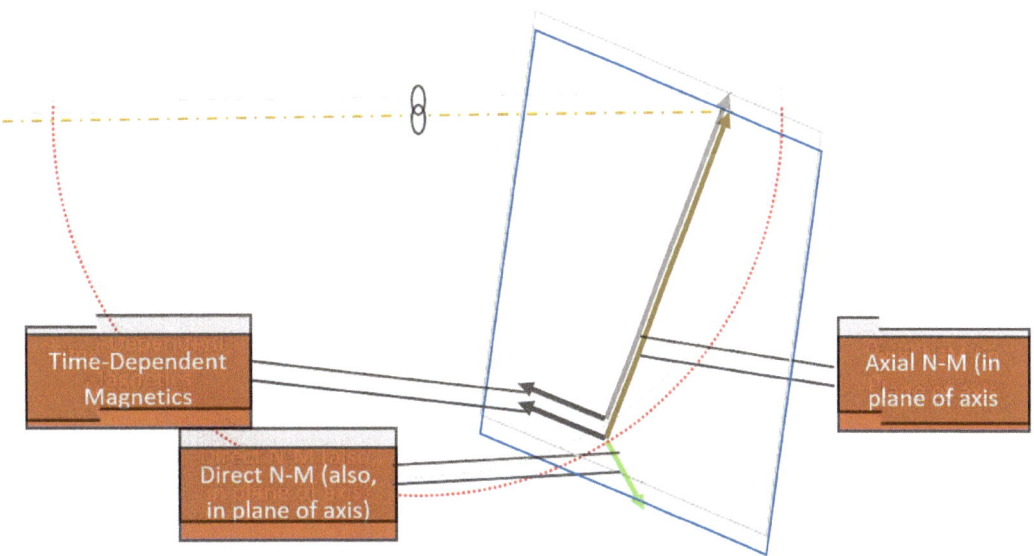

However, one needs to take the position that, given no other particles, an electron movement in that direction will need a field offset (the right-hand rule).

(108)

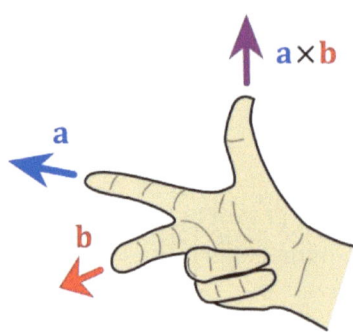

That once you have the 2-vector on the surface of the regular toroid, the counter action is not linear, but at their dot-product (the thumb above or the 'x' out of the paper below). Both the direct nucleostaticmagnetics (strong, red arrow) and the axial nucleostaticmagnetics (weak, brown arrow) are in a plane, and the rules of magnetics force all movements (hashed arrow) to be in that plane.

(109)

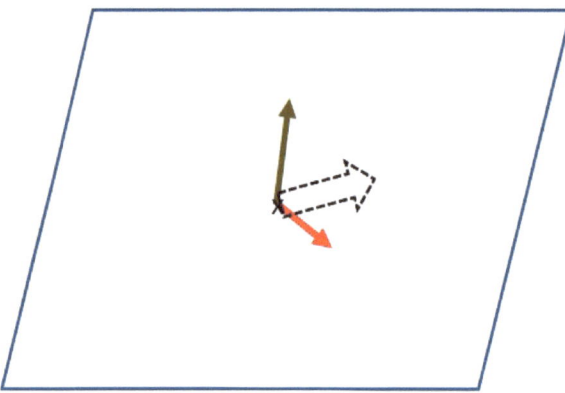

Now, the math gets into matrix algebra as the angle between the red and blue is based upon the $\sin(\theta)$ for the axial (weak), and $\cos(\theta/2)$ for the direct (strong). However, the functions between them then become the tangent (which then cycles that seem 2x full cycles, but a Max(tangent of duopoles) for observation because θ vs $\theta/2$ which both stay in phase, but have that quantum leap reversal for any crossing the equation to the other hemisphere.

The direct nucleostaticmagnetics rotates (strong, with particle movement) at one rate, and the axial nucleostaticmagnetics (weak, with particle movement) rotates as 2x that rate (2x of $\theta/2$ = θ).

(110)

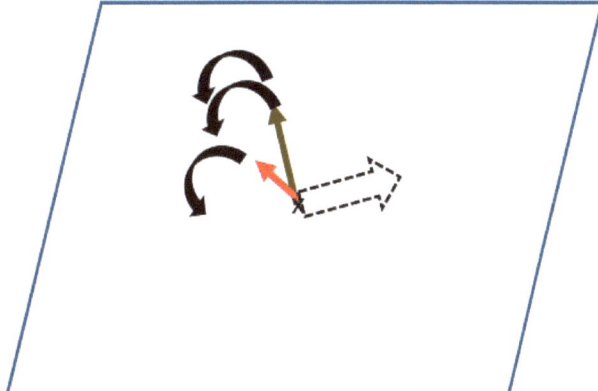

This is the physics-math trigonometry that generates the wave-function of electrons. Further, that really is a clarification of Dirac's 4th Equation that achieves both monopoles and the tangent(θ/2), but as an overlay of the sine and cosine, so it looks like a monopole tangent. It really is the matrix in x,y,z with the z being the magnetic axis in the frame of reference of that particle:

First, the magnitude of the strength nucleostaticmagnetics and the axial nucleostaticmagnetics are the same. That is the application of conservation of energy within a magnetic field (Gauss).

(111)

$$F_{D-N-M} = \frac{M(z,n) * M(e^-)}{d^3} = \frac{M(z,n) * M(e^-)}{d^3} = F_{X-N-M}$$

So, for direct nucleostaticmagnetics (strong) force the direction is particle-to-particle (nucleus to electron) based upon the 'z' a) spherical frame-of-reference and b) θ inclination angle to that axis.

Further, this is **a plane**, not 3D, so we will focus on the combo x,y-direction slice and the z-direction. We will focus on the plane of the electron with the nucleus and its nucleostaticmagnetics axis.

(112)

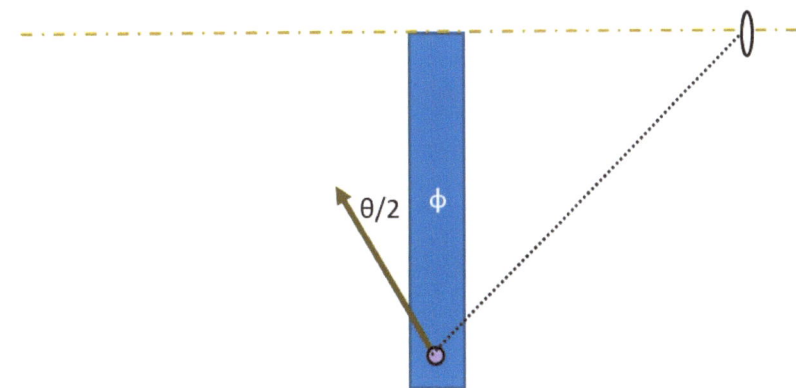

Please note that the **φ plane** (the blue slice that would go out from the paper) of spherical coordinates with 'z' as the axis for inclination angles would not be this same plane. That is different from this plane, and also different from the surface plane. Two complexities (and why we end with 2 x 2). We need to get this slice plane interacting with the surface plane. That is the 2x2 matrix algebra.

Dirac's orientation caused the confusion by isolating the spherical coordinates as the 2nd derivative in the other directions (k_0=0, k_r=0, k_θ=0), so the only direction is the plane with movement along the latitude (k_ϕ).[xvi] He called it φ, but it is not, even in the tangent function that he used. That error in frame-of-reference is the cause of the 100 years of misunderstanding.

Dirac's discuss thinks of this as a slice of a spherical coordinate, and that is the change here. (Please note that is not the tangent plane – which is one of the challenges when we convert to the time-dependent model.) Here is the easy one, for direct nucleostaticmagnetics (strong) force, which is also the same as the electrostatic force). It is based upon that (θ/2);

By that choice, and in all of quantum mechanics, the ending result is in 1 dimension (1D), force strength. So, in that way, even the plane gets eliminated. However, as we use 3D engineering of a physical spinning top where the force is changes in physical rotation of these anisotropic forces, the 1D gets replaces with particle movement behavior in 3D.

So, I have substantially changed the Dirac equation:

- The plane φ substituted with the surface tangent of the toroid surface tangent
- The symmetry and always up-slope of the Dirac tangent function with the 180-degree (and 540 degree) hump with the 270-degree negative infinity to positive discontinuities.

However, we have not changed that equation will give the same results (given quantum methods used today):

- The Force scale at 1/distance-square for 2*r, so 1/distance-cubed
- The discontinuities positions at 90-degrees, 270-degrees, 540-degrees and 630 degrees.
- The full cycle of 720 degrees (if using the hemisphere θ/2 as if that is the θ in the trigonometric function)

So, once again, the good results of quantum mechanics, based upon the unchanged section are not impacted by the physical model that changes the actual equation.

Also, Changing the Plane of the Dirac Equation.

Also, note that the plane is not φ. Instead, the planes are the spinning top dynamic of the direct θ and the axial θ/2:

The electrostatic and direct nucleostaticmagnetics both operate based upon distance between the particles. The leaves those with freedom of moment on the plane of the perpendicular. To move closer or further, then the 1/distance-squared versus 1/distance-cube dynamic occurs.

(113)

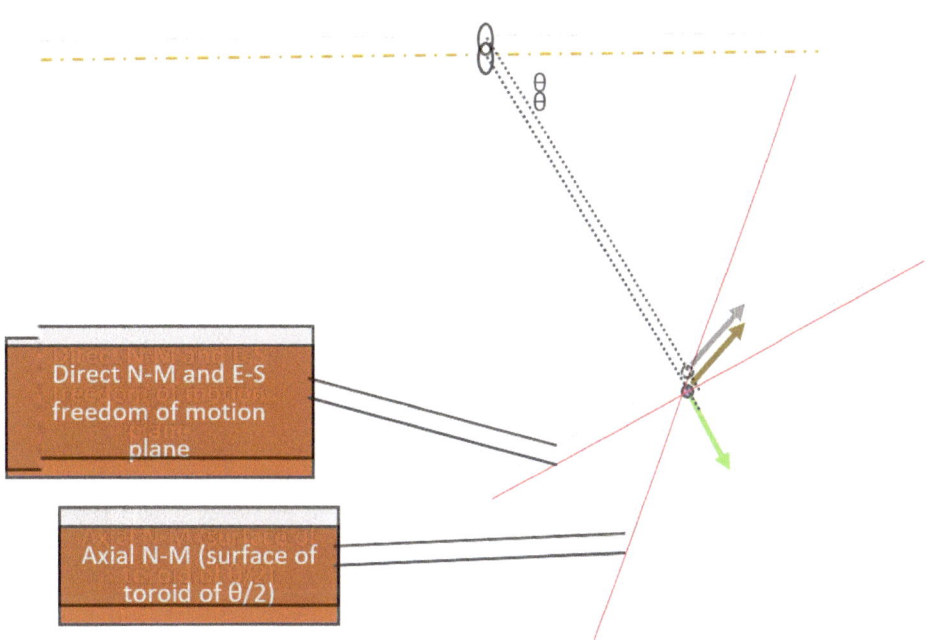

Then you have the spinning top with the axial for (like gravity) creating a progression of particle rotations. For example, the particle rotates on an axis, but the axis progresses around the axial vector:

(114)

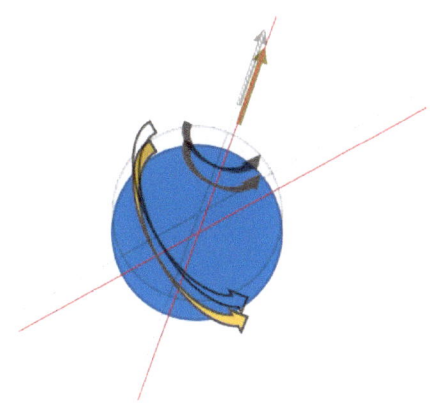

Rotational Forces are Just Sine

The prior chapters spent a lot of effort on the sin*cos function for the calculation of linear strength.

As mentioned, the linear forces are only half of the story; for every linear force, there is rotational force on the other particle-set involved.

For this I repeat the graphic about the four forces for a two-particle interaction. The axis of the 1st particle generates both a linear force on the 2nd (other) particle and b) a rotational force on that axis to align its (1st particle) axis with the direction of that 2nd particle position (only for the strongest as we will discuss in more detail following). Further, the 2nd particle generates both a linear force on the 1st particle and b) a rotational force on its own (2nd particle) axis to align that axis with that 1st particle position:

(115)

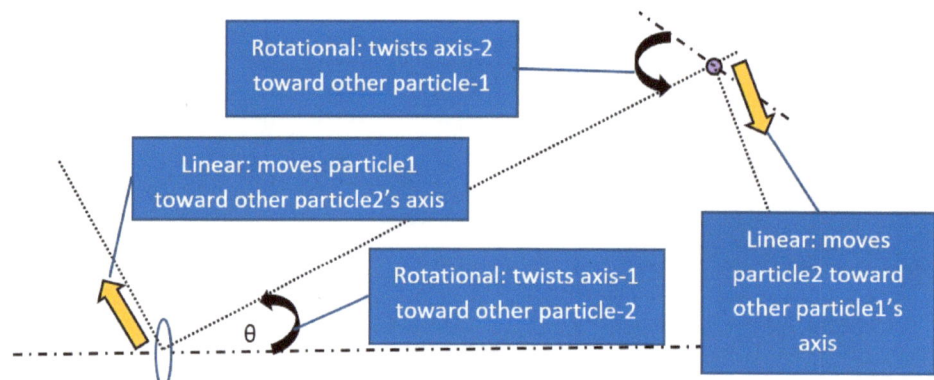

So, I focus on my next paper on applying that logic to get from the above to the Maxwell equations.

Conclusions and Comments

The wording 'nuclear', now changed to nucleostaticmagnetics, had a reason. The forces only apply at distances within the atomic range.

- This is solved by 1/distance-cubed. That way they exist at larger distances, but are immaterial. That that makes the equation correct at any distance – math integrity.

The wording 'strong', now changed to direct nucleostaticmagnetics, had a reason. The forces must be stronger than the huge proton-proton electrostatic repulsion (like-kind repel) at nucleus distances at 1/distance-squared ($1/d^2$).

- This is also solved by 1/distance-cubed ($1/d^3$). The crazy math is that as x > 0, that is x is becoming infinity (so 1/infinity > 0, but the force (1/(1/infinity)-cubed) become infinity-cubed which is greater than the proton-proton electrostatic (like-kind) repulsion at those tiny distances within the nucleus. Infinity-cubed beats infinity squared!

The wording 'weak', now changed to axial nucleostaticmagnetics, had a reason. These forces were a smaller than the 'strong' force.

- This is solved by the sine times cosine (sin*cos) factor for the towards-the-axis force. At the greatest, that factor is 0.500x, but it can go to near zero at the poles or equator, and the poles is the location for most nuclear binding where that is the extra force to keep the amazing nucleus structure of a 3D neutron between every proton (proton-neutron-proton) in place. That towards the axis, keeps the chain at the poles, so strong (direct nucleostaticmagnetics) will do its job keeping the nucleus as a 3D structure.

Quantum mechanics and today's version of quantum theory all start with a method of estimating position and velocity knowing only electrostatic force. However, that is an incomplete 3D engineering, so the results were statistical. However, scientists chose the Copenhagen model that the underlying particles themselves were statistical.

Well, one cannot have a 3D engineering models for probabilities that is based upon only a probability source. One can never escape statistical logic when taking the components as statistical.

It was *chicken-egg*, and the above presents that the chicken (deterministic fundamental forces) generates the egg (ring, flattened spherical field of action for electrons, chemistry, magnetism).

That the next work is re-stating all the magnetic field work from a subatomic static model, with the Maxwell equations applied at every distance consistently.

The herein duopole is a regular toroid by this effect; it is a ring-electron by this effect; it is a disk by this effect. The duopole model generates strange and amazing quantum leaps where forces jump into absolutely different directions.

There are years of work further to re-state every rule of Maxwell and Biot-Savart, but this is a start. Wish me luck. Or, if not, you scientists in the next generation will expand on this work to make every calculation calculated with math integrity.

Big hugs! Let's go understand all the world with common sense, with physics and chemistry as basic mechanics with 3D engineering and math integrity!

Endnotes

[i] Although as presented today, this is a 3-particles system for the nucleus applications of proton-neutron-proton in 3D. That is the current presentation that will get updated by the work here.

[ii] Please note (after you read that section) that that I plan further work that might apply the strength formula over the axis which might create a) sin^2*cos^2 or b) sin^2*cos or c) a different integral of the force over the segment of the axis not balanced by the opposing hemisphere as a viable alternative to simply $sin*cos$. That is almost immaterial, but I need to leave that possibility open for later exploration. Think about it, you have a region left and right, but is that an integral of sin^2 (used here to generate the basic sine function) or sin^3 (if used that would generate the alternatives above). It makes a difference that I have not explored here.

[iii] in the same hemisphere as I shall explain in that section.

[iv] I have 3D computer models.

[v] Please note that the θ/2 is an average. It works at the balancing of E-S and N-M fields which is generally the average electron radius in an atomic model. However, the E-S field strength is usually slightly larger for the outer electrons. In that way, further work is need to determine if the results are averaging (superposition of fields) or 'largest wins'. My work here tends to support that averaging theorem, but I admit the work is thin and incomplete. Think about macro-magnetics, which scientists think of as orthogonal (90 degree) as the tangent plane to the E-S field. Well, at macro-distances, that is by far the strongest field, so it overwhelms that other. I think it really is an average, but the other forces subatomic nucleostaticmagnetics is immaterial if not aggregating multiple molecules. As such, we can use E-S orthogonal for results to many decimal places of precision.

[vi] It is a larger proof to show that the electrostatic and nucleostaticmagnetics forces pass the Newtonian balance. Especially, in 3D over all distances. That is beyond the scope here, but a direction that I plan to pursue later.

[vii] That toroid math from which I suffered myself making that engineering mistake in the many steps to develop this subatomic force and structure model.

[viii] With my clarification that its is a strength average, and that for outer subshell electrons will average towards the E-S field more (as outside the He Bohr radius, the electron-nucleus equilibrium of (ES plus Direct NM) forces and fields is weighted more towards the E-S. The extra repulsions for outer subshells are from inner layer electrons which express a electrostatic field, but at varying angles. That calculation remains beyond the space allowable here.

[ix] Again, see Note v describing that I think this for outer electrons is slightly different (more allocation to the E-S field), but the results is close enough.

[x] So, that is why Dirac chose the θ/2 because that scrunches the graph so the zero passes at 90 degrees (the 2θ position). A tangent(θ/2) has a) the discontinuities, and the b) zero's in the same positions. It works for the things that Dirac could measure.

[xi] Planck.

[xii] Dirac.

[xiii] Built by Arne Vigen using Wolframalpha.com

[xiv] The closest reference is from Biot-Savart, and the root($1+3\cos^2$) comes from that. They applied that for macro-world magnets. And, of course, it applies at the micro-world with math integrity.

[xv] Dirac, P.A.M, 1931. "Quantised Singularities in Electromagnetic Field"

[xvi] The discussion of the latitude is on a toroid or a sphere is the critical issue here. In fact, it is a combination, and that creates the discontinuities of quantum mechanics.

www.ingramcontent.com/pod-product-compliance
Lightning Source LLC
Chambersburg PA
CBHW051914210526
45473CB00006B/2014